THROUGH THE VANISHING POINT
Space in Poetry and Painting

WORLD PERSPECTIVES

Volumes already published

I	Approaches to God	Jacques Maritain
II	Accent on Form	Lancelot Law Whyte
III	Scope of Total Architecture	Walter Gropius
IV	Recovery of Faith	Sarvepalli Radhakrishnan
V	World Indivisible	Konrad Adenauer
VI	Society and Knowledge	V. Gordon Childe
VII	The Transformations of Man	Lewis Mumford
VIII	Man and Materialism	Fred Hoyle
IX	The Art of Loving	Erich Fromm
X	Dynamics of Faith	Paul Tillich
XI	Matter, Mind and Man	Edmund W. Sinnott
XII	Mysticism: Christian and Buddhist	Daisetz Teitaro Suzuki
XIII	Man's Western Quest	Denis de Rougemont
XIV	American Humanism	Howard Mumford Jones
XV	The Meeting of Love and Knowledge	Martin C. D'Arcy, S.J.
XVI	Rich Lands and Poor	Gunnar Myrdal
XVII	Hinduism: Its Meaning for the Liberation of the Spirit	Swami Nikhilananda
XVIII	Can People Learn to Learn?	Brock Chisholm
XIX	Physics and Philosophy	Werner Heisenberg
XX	Art and Reality	Joyce Cary
XXI	Sigmund Freud's Mission	Erich Fromm
XXII	Mirage of Health	René Dubos
XXIII	Issues of Freedom	Herbert J. Muller
XXIV	Humanism	Moses Hadas
XXV	Life: Its Dimensions and Its Bounds	Robert M. MacIver
XXVI	Challenge of Psychical Research	Gardner Murphy
XXVII	Alfred North Whitehead: His Reflections on Man and Nature	Ruth Nanda Anshen
XXVIII	The Age of Nationalism	Hans Kohn
XXIX	Voices of Man	Mario Pei
XXX	New Paths in Biology	Adolf Portmann
XXXI	Myth and Reality	Mircea Eliade
XXXII	History as Art and as Science	H. Stuart Hughes
XXXIII	Realism in Our Time	Georg Lukács
XXXIV	The Meaning of the Twentieth Century	Kenneth E. Boulding
XXXV	On Economic Knowledge	Adolph Lowe
XXXVI	Caliban Reborn	Wilfrid Mellers
XXXVII	Through the Vanishing Point	Marshall McLuhan and Harley Parker

BOARD OF EDITORS
of
WORLD PERSPECTIVES

Sir Kenneth Clark

Richard Courant

Werner Heisenberg

Konrad Lorenz

Robert M. MacIver

Jacques Maritain

Joseph Needham

I. I. Rabi

Sarvepalli Radhakrishnan

Alexander Sachs

C. N. Yang

WORLD PERSPECTIVES · *Volume Thirty-seven*

Planned and Edited by RUTH NANDA ANSHEN

THROUGH THE VANISHING POINT
Space in Poetry and Painting

MARSHALL McLUHAN AND
HARLEY PARKER

1817

HARPER & ROW, PUBLISHERS
New York, Evanston, and London

Grateful acknowledgment is made for permission to reprint excerpts from the following:

"In Memory of W. B. Yeats" by W. H. Auden, from *The Collected Poetry of W. H. Auden* and *Collected Shorter Poems 1927-1957*. Copyright 1940 by W. H. Auden. Reprinted by permission of Random House, Inc., and Faber & Faber Ltd.

"L'Invitation au Voyage" by Charles Baudelaire, from *Paris Spleen*, translated by Louise Varèse. Copyright 1947 by New Directions. Reprinted by permission of New Directions Publishing Corporation.

"The Great Lover" by Rupert Brooke, from *The Collected Poems of Rupert Brooke*. Copyright 1915 by Dodd, Mead & Company. Copyright renewed 1943 by Edward Marsh. Reprinted by permission of Dodd, Mead & Company, Inc.

"Chansons Innocentes" by E. E. Cummings, from *Poems 1923-1954*. Copyright 1923, 1951 by E. E. Cummings. Reprinted by permission of Harcourt, Brace & World, Inc.

"Fragment of an Agon," "A Game of Chess," "Hysteria," *The Waste Land*, and "Whispers of Immortality" by T. S. Eliot, from *Collected Poems 1909-1962*. Copyright 1936 by Harcourt, Brace & World, Inc.; copyright © 1963, 1964 by T. S. Eliot. Reprinted by permission of Harcourt, Brace & World, Inc., and Faber & Faber Ltd.

"Ben Jonson" by T. S. Eliot, from *Selected Essays*. Reprinted by permission of Harcourt, Brace & World, Inc., and Faber & Faber Ltd.

(continued)

THROUGH THE VANISHING POINT: SPACE IN POETRY AND PAINTING. *Copyright © 1968 by Marshall McLuhan and Harley Parker. Printed in the United States of America. All rights reserved. No part of this book may be used or reproduced in any manner whatsoever without written permission except in the case of brief quotations embodied in critical articles and reviews. For information address Harper & Row, Publishers, Incorporated, 49 East 33rd Street, New York, N.Y. 10016.*

I-S

Library of Congress Catalog Card Number: 68-15964
Designed by The Etheredges

"Kafka's Castle Stands Above the World" by Lawrence Ferlinghetti, from *A Coney Island of the Mind.* Copyright © 1958 by Lawrence Ferlinghetti. Reprinted by permission of New Directions Publishing Corporation.

Life with Picasso by Françoise Gilot and Carlton Lake. Copyright © 1964 by McGraw-Hill, Inc. Reprinted by permission of McGraw-Hill Book Company.

The Hidden Dimension by Edward T. Hall. Copyright © 1959 by Edward T. Hall. Reprinted by permission of Doubleday & Company, Inc.

The Popular Arts by Stuart Hall and Paddy Whannel. Copyright © 1964 by Stuart Hall and Paddy Whannel. Reprinted by permission of Pantheon Books, a Division of Random House, Inc., and The Hutchinson Publishing Group Ltd.

"Duns Scotus's Oxford" and "Pied Beauty" by Gerard Manley Hopkins, from *Poems of Gerard Manley Hopkins,* Third Edition, edited by W. H. Gardner. Copyright 1948 by Oxford University Press, Inc. Reprinted by permission of Oxford University Press.

"Loveliest of Trees" by A. E. Housman, from "A Shropshire Lad"— Authorized Edition—from *The Collected Poems of A. E. Housman.* Copyright 1939, 1940, © 1959 by Holt, Rinehart and Winston, Inc. Copyright © 1967 by Robert E. Symons. Reprinted by permission of Holt, Rinehart & Winston, Inc., The Society of Authors, and Jonathan Cape Ltd.

Problematic Man by Gabriel Marcel, translated by Brian Thompson, 1967. Reprinted by permission of Herder & Herder, Inc.

"First Families, Move Over" by Ogden Nash, from *Verses from 1929 On.* Copyright 1935 by Ogden Nash; originally appeared in *The New Yorker.* Reprinted by permission of Little, Brown & Co.

"Hugh Selwyn Mauberley," "In a Station of the Metro" and "Portrait d'une Femme" by Ezra Pound, from *Personae.* Copyright 1926, 1954 by Ezra Pound. Reprinted by permission of New Directions Publishing Corporation.

"Beast of the Sea," "Great Sea," and "The Lands Around My Dwelling" from *Anerca,* edited by E. S. Carpenter, translated by Knud Rasmussen. Copyright © 1959 by Edmund Carpenter. Reprinted by permission of Edmund Carpenter and the heirs of Knud Rasmussen.

"After the Flood" by Arthur Rimbaud, from *Illuminations,* from *Complete Works of Rimbaud,* translated by Wallace Fowlie. Copyright 1966 by The University of Chicago. Reprinted by permission of The University of Chicago Press.

"Esthétique du Mal" by Wallace Stevens, from *Collected Poems of Wallace Stevens.* Copyright 1947 by Wallace Stevens. Reprinted by permission of Alfred A. Knopf, Inc.

"Poem in October" by Dylan Thomas, from *Collected Poems.* Copyright 1946 by Dylan Thomas. Reprinted by permission of New Directions Publishing Corporation, J. M. Dent & Sons Ltd., and the Trustees for the Copyrights of the late Dylan Thomas.

"Byzantium" by William Butler Yeats, from *Collected Poems*. Copyright 1933 by The Macmillan Company, renewed 1961 by Bertha Georgie Yeats. Reprinted by permission of The Macmillan Company, New York, Mr. M. B. Yeats, Macmillan of Canada, and Macmillan & Company Ltd., London.

"An Irish Airman Foresees His Death" by William Butler Yeats, from *Collected Poems*. Copyright 1919 by The Macmillan Company, renewed 1947 by Bertha Georgie Yeats. Reprinted by permission of The Macmillan Company, New York, Mr. M. B. Yeats, Macmillan of Canada, and Macmillan & Company Ltd., London.

"Leda and the Swan" by William Butler Yeats, from *Collected Poems*. Copyright 1928 by The Macmillan Company, renewed 1956 by Georgie Yeats. Reprinted by permission of The Macmillan Company, New York, Mr. M. B. Yeats, Macmillan of Canada, and Macmillan & Company Ltd., London.

"An Introduction to My Plays" and "Symbolism in Painting" by William Butler Yeats, from *Essays and Introductions*. Copyright © 1961 by Mrs. W. B. Yeats. Reprinted by permission of The Macmillan Company, New York.

CONTENTS

WORLD PERSPECTIVES—*What This Series Means*	xv
PREFACE	xxiii
SENSORY MODES	1
TOWARD A SPATIAL DIALOGUE	32
THE EMPEROR'S NEW CLOTHES	237
APPENDICES	
I. A NOTE ON TACTILITY	263
II. A NOTE ON COLOR TV	266

ILLUSTRATIONS

1. *Bison, Altamira Caves, Spain*	34
2. *Chinese Character*	38
3. *The Killing of the Wooers (Scyphos vase)*	42
4. *Scene from the Life of Buddha (relief fragment)*	46
5. *Bronze Crucifixion plaque, St. Johns, County Roscommon, Ireland*	50
St. Mark, from the Gospel Book of the Archbishop of Ebbo of Reims	50
6. *Two Warriors Fighting in a Landscape (Persian manuscript)*	54
7. *Mary Receiving the Announcement of Her Death by Duccio de Buoninsegna*	58
8. *Christ Bearing the Cross (Spanish School, fourteenth century)*	62

9. *February, from the Very Rich Book of Hours of the Duke of Berry by Pol Limbourg*   66
10. *The Journey of the Magi by Sassetta*   72
11. *The Garden of Delights (detail) by Hieronymus Bosch*   76
12. *Vulcan and Aeolus by Piero di Cosimo*   80
13. *The Parable of the Blind by Pieter Brueghel*   84
14. *Persian map of the world (sixteenth century)*   88
15. *The Burial of Count D'Orgaz by El Greco*   92
16. *The Calling of St. Matthew by Michelangelo Caravaggio*   96
17. *Mary Magdalen with a Mirror by Georges de La Tour*   100
18. *Self-Portrait by Rembrandt van Rijn*   104
19. *Portrait of the Countess of Meath by Peter Lely*   108
20. *Perfect Harmony* (L'Accord Parfait) *by Antoine Watteau*   112
21. *Ceiling fresco of the Kaisersaal by Giovanni Battista Tiepolo*   116
22. *Landscape with a Bridge by Thomas Gainsborough*   120
23. *Cows at Derwentwater by Chiang Yee*   124
24. *Iron Works at Coalbrookdale by Pierre Jacques de Loutherbourg*   128
25. *The Death of Wolfe by Benjamin West*   132
26. *The Nightmare by Henri Füssli*   136
27. *"The Tyger" (page from the original edition of* Songs of Experience*) by William Blake*   140
28. *Odalisque by Jean-Auguste Dominique Ingres*   144
29. *Sir Galahad by Dante Gabriel Rossetti*   148
30. *A Hill-Side with Tramps Reposing by George Morland*   152
31. *Steamer in a Snowstorm by Joseph Mallord William Turner*   156
32. *Arab Rider Attacked by Lion by Eugène Delacroix*   160
33. *Facsimile page from the original manuscript of* Alice's Adventures Under Ground *by Lewis Carroll*   162
    Page from Alice's Adventures in Wonderland *by Lewis Carroll, illustrated by John Tenniel*   164

34.	*Arrangement in Black and Gray (The Artist's Mother) by James Abbott McNeill Whistler*	*168*
35.	*The Dream by Henri Rousseau*	*172*
36.	*The Scream by Edvard Munch*	*176*
37.	*A Sunday Afternoon on the Island of La Grande Jatte by Georges Seurat*	*180*
38.	*Intrigue by James Ensor*	*184*
39.	*The Twittering Machine by Paul Klee*	*188*
40.	*Portrait of an Englishwoman by Wyndham Lewis*	*192*
41.	*I and the Village by Marc Chagall*	*196*
42.	*Drawing by Saul Steinberg*	*200*
43.	*Swan Upping by Stanley Spencer*	*204*
44.	*Still Life: The Table by Georges Braque*	*208*
45.	*1967 Advertisement*	*212*
46.	*The Trip by Harley Parker*	*216*
47.	*Full Fathom Five by Jackson Pollock*	*220*
48.	*I Am That I Am by Fay Zetlin*	*230*
49.	*In Bedlam (Scene VIII from The Rake's Progress) by William Hogarth*	*234*

WORLD PERSPECTIVES

What This Series Means

It is the thesis of *World Perspectives* that man is in the process of developing a new consciousness which, in spite of his apparent spiritual and moral captivity, can eventually lift the human race above and beyond the fear, ignorance, and isolation which beset it today. It is to this nascent consciousness, to this concept of man born out of a universe perceived through a fresh vision of reality, that *World Perspectives* is dedicated.

Man has entered a new era of evolutionary history, one in which rapid change is a dominant consequence. He is contending with a fundamental change, since he has intervened in the evolutionary process. He must now better appreciate this fact and then develop the wisdom to direct the process toward his fulfillment rather than toward his destruction. As he learns to apply his understanding of the physical world for practical purposes, he is, in reality, extending his innate capacity and augmenting his ability and his need to communicate as well as his ability to think and to create. And as a result, he is substituting a goal-directed evolutionary process in his struggle against environmental hardship for the slow, but effective, biological evolution which produced modern man through mutation and natural selection. By intelligent intervention in the evolutionary process man has greatly accelerated and greatly expanded the range of his possibilities. But he has not changed the basic fact that it remains a trial and error process, with the danger of taking paths that lead to sterility of mind and heart, moral apathy and intellectual inertia; and even producing social dinosaurs unfit to live in an evolving world.

Only those spiritual and intellectual leaders of our epoch who have a paternity in this extension of man's horizons are invited to participate in this Series: those who are aware of the truth that beyond the divisiveness among men there exists a primordial unitive power since we are all bound together by a common humanity more fundamental than any unity of dogma; those who recognize that the centrifugal force which has scattered and atomized mankind must be replaced by an integrating structure and process capable of bestowing meaning and purpose on existence; those who realize that science itself, when not inhibited by the limitations of its own methodology, when chastened and humbled, commits man to an indeterminate range of yet undreamed consequences that may flow from it.

This Series endeavors to point to a reality of which scientific theory has revealed only one aspect. It is the commitment to this reality that lends universal intent to a scientist's most original and solitary thought. By acknowledging this frankly we shall restore science to the great family of human aspirations by which men hope to fulfill themselves in the world community as thinking and sentient beings. For our problem is to discover a principle of differentiation and yet relationship lucid enough to justify and to purify scientific, philosophic and all other knowledge, both discursive and intuitive, by accepting their interdependence. This is the crisis in consciousness made articulate through the crisis in science. This is the new awakening.

Each volume presents the thought and belief of its author and points to the way in which religion, philosophy, art, science, economics, politics and history may constitute that form of human activity which takes the fullest and most precise account of variousness, possibility, complexity and difficulty. Thus *World Perspectives* endeavors to define that ecumenical power of the mind and heart which enables man through his mysterious greatness to re-create his life.

This Series is committed to a re-examination of all those sides of human endeavor which the specialist was taught to believe he could safely leave aside. It attempts to show the structural kinship between subject and object; the indwelling of the one in the other. It interprets present and past events impinging on human life in our growing World Age and envisages what man may yet attain when summoned by an unbending inner necessity to the quest of what is most exalted in him. Its purpose is to offer new vistas in terms of world and human development while refusing to betray the intimate correlation between universality and individuality, dynamics and form, freedom and destiny. Each author deals with the increasing realization that spirit and nature are not separate and apart; that intuition and reason must regain their importance as the means of perceiving and fusing inner being with outer reality.

World Perspectives endeavors to show that the conception of wholeness, unity, organism is a higher and more concrete conception than that of matter and energy. Thus an enlarged meaning of life, of biology, not as it is revealed in the test tube of the laboratory but as it is experienced within the organism of life itself, is attempted in this Series. For the principle of life consists in the tension which connects spirit with the realm of matter, symbiotically joined. The element of life is dominant in the very texture of nature, thus rendering life, biology, a transempirical science. The laws of life have their origin beyond their mere physical manifestations and compel us to consider their spiritual source. In fact, the widening of the conceptual framework has not only served to restore order within the respective branches of knowledge, but has also disclosed analogies in man's position regarding the analysis and synthesis of experience in apparently separated domains of knowledge, suggesting the possibility of an ever more embracing objective description of the meaning of life.

Knowledge, it is shown in these books, no longer consists in a manipulation of man and nature as opposite forces, nor in the reduction of data to mere statistical order, but is a means of liberating mankind from the destructive power of fear, pointing the way toward the goal of the rehabilitation of the human will and the rebirth of faith and confidence in the human person. The works published also endeavor to reveal that the cry for patterns, systems and authorities is growing less insistent as the desire grows stronger in both East and West for the recovery of a dignity, integrity and self-realization which are the inalienable rights of man who may now guide change by means of conscious purpose in the light of rational experience.

The volumes in this Series endeavor to demonstrate that only in a society in which awareness of the problems of science exists can its discoveries start great waves of change in human culture, and in such a manner that these discoveries may deepen and not erode the sense of universal human community. The differences in the disciplines, their epistemological exclusiveness, the variety of historical experiences, the differences of traditions, of cultures, of languages, of the arts, should be protected and preserved. But the interrelationship and unity of the whole should at the same time be accepted.

The authors of *World Perspectives* are of course aware that the ultimate answers to the hopes and fears which pervade modern society rest on the moral fibre of man, and on the wisdom and responsibility of those who promote the course of its development. But moral decisions cannot dispense with an insight into the interplay of the objective elements which offer and limit the choices made. Therefore an understanding of what the issues are, though not a sufficient condition, is a necessary prerequisite for directing action toward constructive solutions.

Other vital questions explored relate to problems of international understanding as well as to problems dealing with prejudice and the resultant tensions and antagonisms. The growing perception and responsibility of our World Age point to the new reality that the individual person and the collective person supplement and integrate each other; that the thrall of totalitarianism of both left and right has been shaken in the universal desire to recapture the authority of truth and human totality. Mankind can finally place its trust not in a proletarian authoritarianism, not in a secularized humanism, both of which have betrayed the spiritual property right of history, but in a sacramental brotherhood and in the unity of knowledge. This new consciousness has created a widening of human horizons beyond every parochialism, and a revolution in human thought comparable to the basic assumption, among the ancient Greeks, of the sovereignty of reason; corresponding to the great effulgence of the moral conscience articulated by the Hebrew prophets; analogous to the fundamental assertions of Christianity; or to the beginning of the new scientific era, the era of the science of dynamics, the experimental foundations of which were laid by Galileo in the Renaissance.

An important effort of this Series is to re-examine the contradictory meanings and applications which are given today to such terms as democracy, freedom, justice, love, peace, brotherhood and God. The purpose of such inquiries is to clear the way for the foundation of a genuine *world* history not in terms of nation or race or culture but in terms of man in relation to God, to himself, his fellow man and the universe, that reach beyond immediate self-interest. For the meaning of the World Age consists in respecting man's hopes and dreams which lead to a deeper understanding of the basic values of all peoples.

World Perspectives is planned to gain insight into the meaning of man, who not only is determined by history but who also determines history. History is to be understood as concerned not only with the life of man on this planet but as including also such cosmic influences as interpenetrate our human world. This generation is discovering that history does not conform to the social optimism of modern civilization and that the organization of human communities and the establishment of freedom and peace are not only intellectual achievements but spiritual and moral achievements as well, demanding a cherishing of the wholeness of human personality, the "unmediated wholeness of feeling and thought," and constituting a never-ending challenge to man, emerging from the abyss of meaninglessness and suffering, to be renewed and replenished in the totality of his life.

Justice itself, which has been "in a state of pilgrimage and crucifixion" and now is being slowly liberated from the grip of social and political demonologies in the East as well as in the West, begins to question its own premises. The modern revolutionary movements which have challenged the sacred institutions of society by protecting social injustice in the name of social justice are here examined and re-evaluated.

In the light of this, we have no choice but to admit that the *un*-freedom against which freedom is measured must be retained with it, namely, that the aspect of truth out of which the night view appears to emerge, the darkness of our time, is as little abandonable as is man's subjective advance. Thus the two sources of man's consciousness are inseparable, not as dead but as living and complementary, an aspect of that "principle of complementarity" through which Niels Bohr has sought to unite the quantum and the wave, both of which constitute the very fabric of life's radiant energy.

There is in mankind today a counterforce to the sterility and danger of a quantitative, anonymous mass culture; a new, if sometimes imperceptible, spiritual sense of convergence toward human and world unity on the basis of the sacredness of each human person and respect for the plurality of cultures. There is a growing awareness that equality may not be evaluated in mere numerical terms but is proportionate and analogical in its reality. For when equality is equated with interchangeability, individuality is negated and the human person extinguished.

We stand at the brink of an age of a world in which human life presses forward to actualize new forms. The false separation of man and nature, of time and space, of freedom and security, is acknowledged, and we are faced with a new vision of man in his organic unity and of history offering a richness and diversity of quality and majesty of scope hitherto unprecedented. In relating the accumulated wisdom of man's spirit to the new reality of the World Age, in articulating its thought and belief, *World Perspectives* seeks to encourage a renaissance of hope in society and of pride in man's decision as to what his destiny will be.

World Perspectives is committed to the recognition that all great changes are preceded by a vigorous intellectual re-evaluation and reorganization. Our authors are aware that the sin of *hubris* may be avoided by showing that the creative process itself is not a free activity if by free we mean arbitrary, or unrelated to cosmic law. For the creative process in the human mind, the developmental process in organic nature and the basic laws of the inorganic realm may be but varied expressions of a universal formative process. Thus *World Perspectives* hopes to show that although the present apocalyptic period is one of exceptional tensions, there is also at work an exceptional movement toward a compensating unity which refuses to violate the ultimate moral power at work in the universe, that very power upon which all human effort must at last depend. In this way we may come to understand that there exists an inherent independence of spiritual and mental growth which, though conditioned by circumstances, is never determined by circumstances. In this way the great plethora of human knowledge may be correlated with an insight into the nature of human nature by being attuned to the wide and deep range of human thought and human experience.

In spite of the infinite obligation of men and in spite of their finite power, in spite of the intransigence of nationalisms, and in spite of the homelessness of moral passions rendered ineffectual by the scientific outlook, beneath the apparent turmoil and upheaval of the present, and out of the transformations of this dynamic period with the unfolding of a world-consciousness, the purpose of *World Perspectives* is to help quicken the "unshaken heart of well-rounded truth" and interpret the significant elements of the World Age now taking shape out of the core of that undimmed continuity of the creative process which restores man to mankind while deepening and enhancing his communion with the universe.

RUTH NANDA ANSHEN

PREFACE

Since the advent of electric circuitry in the early nineteenth century, the need for sensory awareness has become more acute. Perhaps the mere speed-up of human events and the resulting increase of interfaces among all men and institutions insure a multitude of innovations that upset all existing arrangements whatever.

By the same token, men have groped toward the arts in hope of increased sensory awareness. The artist has the power to discern the current environment created by the latest technology. Ordinary human instinct causes people to recoil from these new environments and to rely on the rear-view mirror as a kind of repeat or *ricorso* of the preceding environment, thus insuring total disorientation at all times. It is not that there is anything wrong with the old environment, but it simply will not serve as navigational guide to the new one.

Paradoxically, war as an educational institution serves to bring people into contact with the new technological environments that the artist had seen much earlier. Complementarily, education can be seen as a kind of war conducted by the Establishment to keep the sensory life in line with existing commitments. It also serves to keep the sensory life out of touch with innovation. "History is a nightmare from which I am trying to awake."

If war can become a form of education, art ceases to be a form of self-expression in the electric age. Indeed, it becomes a necessary kind of research and probing. Ashley Montagu has pointed out that the more civilization, the more violence. What he fails to note is the reason for this.

Civilization is founded upon the isolation and domination of society by the visual sense. The visual sense creates a kind of human identity that is extremely fragmented. To retain such an image of the self requires persistent violence, both to one's self and to others. As Joyce put it, "Love thy label as thyself." Labels as classification are extreme forms of visual culture. As the visual bias declines, the other senses come into play once more. The arts have been expounding this fact for more than a century.

M.M.
H.P.

". . . the flat earth Blake associated with ordinary imaginative possibility is replaced by the spectral ball of universes seen through the wrong or dwarfing end of vision, the inverted vortex of Newtonian observation."

HAROLD BLOOM, *Blake's Apocalypse*

SENSORY MODES

As the Western world has invested every aspect of its waking life with visual order, with procedures and spaces that are uniform, continuous and connected, it has progressively alienated itself from needful involvement in its subconscious life.

Our manner of juxtaposing a poem with a painting is designed to illuminate the world of verbal space through an understanding of spaces as they have been defined and explored through the plastic arts. The verbal medium is so completely environmental as to escape all perceptual study in terms of its plastic values. Everybody can talk, but few can paint. A dialogue between the different forms and qualities of the sister arts of poetry and painting needs no defense, but there has been little exercise of such dialogue, especially with particular references. There has been some speculation from time to time on the lines of *ut pictura poesis*. The advantage of using two arts, both poetry and painting, simultaneously is that one permits a journey inward, the other a journey outward to the appearance of things. The continuity of interface and dialogue between the sister arts should provide a rich means of training perception and sensibility. We hope that our readers will be moved to inundate us with new suggestions of how to exploit this approach. We have resisted essay-like presentations: we hope that the jagged edges of our iconic thrusts and queries will serve to open up, rather than to enclose, the imagery of the perceptual field.

In many of the sections of the book the reader will encounter a concern with the differences between iconic and illustrative modes in art and poetry. It is our purpose to provide a contemporary audience with the tools for discovery of a common ground among the manifestations of art in the world. Though the artistic intentions of the primitive artist and the Renaissance artist may be poles apart, the artistic *effect* under all conditions is a situation that serves to heighten perception. All the arts might be considered to act as counterenvironments or countergradients. Any environmental form whatsoever saturates perception so that its own character is imperceptible; it has the power to distort or deflect human awareness. Even the most popular arts can serve to increase the level of awareness, at least until they become entirely environmental and unperceived.

THE RETURN OF FORMAL SPACE

Our new awareness of territoriality: the return of formal space.

In its August 26, 1966, issue, *Life* magazine featured the studies of Robert Ardrey. Ardrey had first mentioned territoriality in his *African Genesis,* when he explained how the conditions of the nineteenth-century zoo had deflected attention from a prime fact of animal behavior—namely, the animal's need to define and patrol a space of its own. This space is called into being by sounds, by odors, by colors—in short, by that orchestration of the senses compatible with the total life of the species.

The conventional zoo, a rationally and visually contrived space, not only confirmed an unconscious pictorial bias of scientists and spectators, but also eliminated the complex spaces generated by the animals in their normal habitat. The current rediscovery of territorial space points dramatically to the changing sensibility and space orientation of the population in the electronic age. Visual values have lost their power to obliterate the boundaries and forms of space proper to our other senses.

One of the best leads we have to our changing concepts of space probably lies in a study of the mores and patterns of our children.

THE LACK OF HUMAN SPACE RECEPTORS

Although *The Child's Conception of Space* by Piaget and Inhelder explores the changing spatial experience of children, there has so far been no guide to the changing spatial experience that adults typically encounter in poetry and painting. This is understandable when we consider that to contemporary man space is a cliché, an unexamined assumption; it is environmental, and modern man is therefore unaware of it. Interest in space in painting has been mainly in the area of the three-dimensional illusion. The work of Edward T. Hall in *The Silent Language* reminds us that the spatial relations between figures in life and art remain relatively unstudied. Hall develops this theme further in *The Hidden Dimension* (New York, Doubleday & Company, Inc., 1959):

... People brought up in different cultures learn as children, without ever knowing that they have done so, to screen out one type of information while paying close attention to another. Once set, these perceptual patterns apparently remain quite stable throughout life. The Japanese, for example, screen visually in a variety of ways but are perfectly content with paper walls as acoustic screens.

Spending the night at a Japanese inn while a party is going on next door is a new sensory experience for the Westerner. In contrast, the Germans and the Dutch depend on thick walls and double doors to screen sound, and have difficulty if they must rely on their own powers of concentration to screen out sound. If two rooms are the same size and one screens out sound but the other doesn't, the sensitive German who is trying to concentrate will feel less crowded in the former because he feels less intruded on.

In the use of the olfactory apparatus Americans are culturally underdeveloped. The extensive use of deodorants and the suppression of odor in public places results in a land of olfactory blandness and sameness that would be difficult to duplicate anywhere else in the world. This blandness makes for undifferentiated spaces and deprives us of richness and variety in our life. It also obscures memories, because smell evokes much deeper memories than either vision or sound.

In *Psychopathology and Education of the Brain-Injured Child,* Strauss states:

> There is nothing innate in the human nervous system which gives us direct information concerning space. There is no specialized space receptor. Projections of images into a space world are the result of careful focalization of certain rather subtle cues and, as such, is a learned phenomenon. It would seem further that none of the many clues which we use to locate ourselves in space is sufficient in itself. Any one clue is subject to the effects of certain types of interference which, if we depended upon this clue alone, would give us an incomplete picture. It is the interrelationship of these many clues which gives us a clear and well structured space world.

In fact, this interrelationship is seldom accomplished with any degree of efficacy for the reasons observed by Eric Bentley in *The Life of the Drama* (New York, Atheneum, 1965):

> Perception is riveted to need. Our real needs being relatively few, our perceptions are relatively few. They are also relatively faint and incomplete and inaccurate.

Bentley further observes:

> All too often we do not see; we do not look; we have preconceptions. Casting a hasty, nervous glance in front, to make sure

that we don't actually collide with anything, we fit together what we half see with what we assume we know.

In F. C. Bartlett's *Remembering,* there is extensive illustration of the principle that perception itself is a kind of remembering. Almost in the moment of perception, a simultaneous afterimage or effect occurs in the subconscious. All sensation, Dallas Smyth has pointed out (in *The Problem of Perception*), is always 100 percent. But the afterimage which we provide to complement such sensation is an altogether different case. Only the artist has the power to elevate it to the conscious life. Our response to experience is typically so inadequate and confused that without the artistic confrontation of the unconscious image the human condition becomes confused indeed.

INDIFFERENCE OF NATURE TO THE IMAGINATIVE LIFE

According to Adolf Hildebrand (in *The Interpretation of Art* by Solomon Fishman, University of California Press, 1963):

> Nature is heartlessly indifferent to the needs of the imaginative life. . . . We have no guarantee that in nature the emotional elements will be combined appropriately with the demands of the imaginative life, and it is, I think, the great occupation of the graphic arts to give us first of all order and variety in the sensuous plane and then so to arrange the sensuous presentment of objects that the emotional elements are elicited with an order and appropriateness altogether beyond what Nature herself provides.

Hildebrand's *Problem of Form* (1893) had extensive influence in getting attention for the action of sensory completion as the accompaniment of all perception. It was this necessary "closure" or sensory completion that yielded "significant form" as compared with ordinary uncompleted experience of forms.

The great advantage of studying space in the arts is that the arts offer an extraordinary range of sensory situations for the training of perception. For example, it is not commonly understood that the visual is the only sense which creates the illusion of uniform, connected spaces. The man who lives in an aural world lives at the center of a communications sphere, and he is bombarded with sensory data from all sides simultaneously. The aurally structured culture has none of the tracts of visual space long regarded as "normal," "natural" space by literate societies. The painter who works within the confines of a visual, literate culture has to cope with a milieu in which all spaces tend to be connected. It is a world of logic and story lines.

In *Art and Illusion,* Gombrich comments:

> After the many weighty tomes that have been written on how space is rendered in art, Steinberg's trick drawings serve as a welcome reminder that it is never space which is represented but familiar things in situations.

It might seem to the casual reader that Gombrich assumes that the only real space is based on vision. This is to postulate that space is a container for things. On the contrary, as painters well know, space is created or evoked by all manner of associations among colors, textures, sounds and their intervals. The works of painters like Jackson Pollock, for example, present spaces evoked by proprioception and kinesis, as well as other sensory relationships.

WHY THE BALINESE SAY "WE HAVE NO ART"

By beginning our probings of the varieties of space with cave paintings, we can more easily illuminate the concept that there is a parallel between preliterate and postliterate cultures. The primitive lived in a world in which all knowledge and skill were simultaneously accessible to all members of the group; contemporary man has created an information environment that embraces all technologies and all cultures in an inclusive experience.

The Balinese, who have no word for art, say, "We do everything as well as possible." This delightful observation draws attention to the fact that primitive art serves quite a different end from Western art. Like the Balinese, however, Electronic Man approaches the condition in which it is possible to deal with the entire environment as a work of art. This presents no solution to the previous problem of decorating the environment. Quite the contrary. The new possibility demands total understanding of the artistic function in society. It will no longer be possible merely to add art to the environment.

BETWEEN THE WHEEL AND THE CIRCUIT

Between Neolithic and Electronic Man there have not been many new technologies, but much application of power to old ones. Since the planter succeeded the hunter, since man began to specialize in arts and crafts and husbandry, cultures have been formed on single gradients moving toward specific ends and destinies. But Man the Hunter, Paleolithic Man as he is called, lived like us in a kind of "zero gradient." He did not specialize but took his total environment as the world in which to acquire perceptual skill. Today, in the age of electric circuitry, when information retrieval can be both instant and total, the intervening ages of specialism between us and Paleolithic Man the Hunter seem quaint and odd.

Electronic Man also has to train his perceptions in relation to a total environment that includes all previous cultures. Home *is* the hunter—at least so say the Nielson audience-rating agencies. In Joyce's *Finnegans Wake* we read: "Though he might have been more humble, there's NO POLICE LIKE HOLMES." The modern sleuth is unmistakably the all-round hunter.

By rehearsing the process of the crime, the sleuth provides a ritualistic replay or *ricorso* that acts both as revelation and purgation of the original crime. Edgar Allan Poe perceived this when he made his detective Dupin an aesthete.

The present study strives to guide the reader through the sensory mazes evoked by technologies old and new and to explain why, in terms of spatial form in poetry and painting, the Medieval and primitive worlds have so much in common with modern experience.

8 THE DISCOVERY OF DISCONTINUITY IN MEDIEVAL ART AND LETTERS
SUPERIOR SOPHISTICATION OF THE MEDIEVAL EDITOR

Much of *A Preface to Chaucer* by D. W. Robertson (Princeton University Press, 1962) is concerned with explaining how the world of space and time in the art of Chaucer is discontinuous and multileveled. For the modern scholar, the discovery of discontinuity creates dismay since it disrupts his ordinary procedures and classifications. Robertson, for example, declares that the principles of conventional philology are quite inadequate to the task of establishing an encounter with the Medieval habits of multileveled exegesis:

> The helplessness of scientific philology as it is now conceived before allegory is very easy to illustrate even where the simplest form of allegory, the trope described by medieval grammarians, is concerned. This trope consists, as Isidore of Seville describes it, of "alieniloquium." One thing is said by the words, but something else is understood. There are, he says, numerous forms which the figure may take, but seven may be singled out as being most important. They are irony (deriding through praise), antiphrasis, aenigma, charientismos, paroemia (proverbial expression), sarcasm, and astysmos (sarcasm without bitterness). Surely, these tropes should be more evident to modern sophistication than they were to our plodding medieval ancestors. But they are not. For example, modern critics of Lucan almost without exception regard the invocation to Nero in the "Pharsalia" as a panegyric, but medieval commentators from the tenth century to the Renaissance treat the dedication as being ironic.

Robertson makes the point that the representational aspects in the cinerary urn are irrelevant inasmuch as the idea, rather than interaction of the represented figures, is expected to evoke the response. With an Etruscan urn the spectator is not supposed to be viewing the spaces depicted. Robertson wants us to see that the empathy or involvement of the viewer is multisensuous or conceptual rather than fragmented and visual. (See E. H. Gombrich, *Art and Illusion*, Pantheon Books, 1961.) It was only with the fantasy of a tale like *Alice in Wonderland* that nineteenth-century man could make an entree into such diverse spaces. A modern art gallery invites the viewer to be as adventurous and daring as an astronaut in probing new modalities of space. Even the virtuosity of Medieval multileveled space fails to daunt the spectator of today. It was recently pointed out that Gemini V astronauts were complaining of overwork. The viewer of contemporary art finds himself in a similar position. Distortion of the El Greco kind, for example, is produced by a use of multiple visual spaces to exhibit the same situation. The gargoyle, on the other hand, uses spaces generated by the interplay of all the senses.

Georges Poulet in his *Studies in Human Time* discusses the great differences between Medieval and Renaissance response to the experience of time: "For the man of the Middle Ages, then, there was not one duration only. There were *durations*, ranked one above another, and not only in the universality of the exterior world but within himself, in his own nature, in his own human existence."

Significantly, Poulet assumes that the world of space is uniform as if it presented a uniform and continuous character to all men. It is, however, central to the understanding of space in poetry and painting that we recognize the same cultural diversity of space as of time. Although Poulet is acutely aware of how to discriminate among temporal phases, it has apparently not occurred to him that similar discrimination is needed in our approach to spatial experience.

While the chicken may or may not be the egg's idea of getting more eggs, it must be understood that discontinuous times and spaces are always coexistent. One without the other is impossible, just as continuous time and space always go in conjunction. Our perception of both times and spaces is *learned*.

The same culture will impose the mode of making both time and space on all its members.

THE VISUAL GRADIENT AND THE RISE OF FRAGMENTATION

Again Poulet observes that in the seventeenth century "Human thought no longer feels itself as part of things." The steady increase of visual culture contributed directly to this sense of alienation. Thought distinguishes itself from things in order to reflect upon them. Hamlet and Bosch are only two of the numerous Renaissance figures who proclaim their sense of alienation in the new world of intense visual stress and fragmentation. One of the paradoxes of intense stress on visual experience is that it results in fragmentation. It is paradoxical because the visual mode, of itself, in isolation, engenders a space which is uniform and continuous and connected. The reason that high visual stress leads to fragmentation of experience seems to be that sight has the unique power to separate or to capture *single aspects* of space in brief moments of time. The other senses cannot duplicate this feat. Andrew Marvell alludes to the divergent characters of time and space in "To His Coy Mistress":

> But at my back I always hear
> Time's wingèd chariot hurrying near;
> And yonder all before us lie
> Deserts of vast eternity.

Time he can hear, and space he can see; but the entire poem develops these conflicting modes of auditory and visual space. A main trend of science in the seventeenth century was the pursuit of measurement and observation by analysis and fragmentation.

THE SHAKESPEAREAN STRATEGY

The seventeenth-century interassociation of varied levels of fact had begun to appear incompatible with the new science of the time. Bishop Sprat in his *History of the Royal Society* notes the need for desperate remedies in matters of speech and expression:

> They have therefore been most rigorous in putting in execution the only Remedy that can be found for this extravagance; and that has been a constant resolution, to reject all the amplifications, digressions, and swellings of style; to return back to the primitive purity, and shortness, when men deliver'd so many things, almost in an equal number of words. They have exacted from all their members, a close, naked, natural way of speaking, positive expressions, clear senses; a native easiness; bringing all things as near the Mathematical plainess, as they can; and preferring the language of Artizans, Countrymen, and Merchants, before that, of Wits, or Scholars.

Much in the way in which a perspective painter sought to isolate a particular moment and a particular space, Bishop Sprat dreamed that the members of the Royal Society could achieve a similar specialist virtuosity by adhering to these virtues of simple and lucid prose.

Patrick Cruttwell, in *The Shakespearean Moment,* has discussed the course of the Renaissance fragmentation of culture: "The sonnets, then, give us the perfect text through which to see what really happened to the minds of men in this crucial decade, and especially to the poetry which expressed those happenings." This is the decade of the 1590's when "the blue-eyed lady of Spenser and Botticelli, the imaginary lady of imaginary nights whose behaviour was as predictable as her looks," was thrown overboard with a great deal of late Medieval and early Renaissance impedimenta.

If Hamlet is an alienated man isolated from his society, his gaze riveted on the rear-view mirror, the hero, Fortinbras—or the decisive man of action, the well-adjusted man—appears to the Hamlets as a being who is prepared to throw his whole manhood away for the most momentary and splintered of reasons and causes. The man of action finds honor "in a straw"; he dies "for a fantasy, a trick of fame." Cruttwell sees Shakespeare as the only poet to have grappled centrally with this problem of disintegration, and not only to have confronted it, but also to have resolved it.

In his *Elizabethan Poetry* Hallett Smith describes the technique by which Shakespeare achieved this unity amidst disintegration:

> ... Shakespeare, seizing upon the sonnet fashion perhaps because it gave him more immediate opportunities than the drama, developed out of the variety, emotional analysis, and passion of the Petrarchan mode a more serious and profound use of metaphor. In his achievement lies the most curious and the most valuable answer to the Elizabethan problem in this exotic form; how to record experience and the analysis of it simultaneously; how to achieve a combination of probability and strangeness; how to make the sonnet cycle seem to reflect actual life instead of the stiff and outworn situation of the distant lady and the despondent wooer, but to reflect life not so much in its external conditions as in its inward meaning and significance.

A man's reach must exceed his grasp or what's a metaphor?

The Renaissance was unconsciously engaged in creating a pervasive visual space that was uniform, continuous and connected, but the Middle Ages had had a very different kind of space as its psychic and social environment. One of the reasons for this was the "idea" behind Medieval representation. It was the idea rather than the psychological narrative connectives between figures that was central to Medieval communication. Thus there was no need for a "rational" or continuous space in which the figures could find psychological interaction. At one Renaissance extreme, the spectator finds it easy to put himself into the painting as if his space and the space in the painting were the same. This became possible when the "vanishing point" was established "inside" the painting. With the vanishing point came the illusion that space was a continuum between spectator and art situation. The spectator becomes part of the lines of force which find their focus in the vanishing point. In Medieval painting, on the other hand, the focus and vanishing point are *in* the spectator. In contrast, in the seventeenth century, when a portrait turns its eye *on* the observer it creates a dualism that is intended to be noticed. The portrait becomes, in effect, a self-portrait in which the subject is also the observer of the painting. The painting becomes a mirror with, as it were, a psychological vanishing point in the viewer. Here we encounter the world of Descartes and of Hamlet. This use of art as a mirror proved to be an original Renaissance method of involving the audience in a reaction and by making the audience the actor. For the first time in the history of art the spectator shared the point of view of the artist. There would seem to be here some slight anticipation of modern painting: a sharing in the creative process by the spectator.

Perspective itself is a mode of perception which in its very nature moves toward specialism and fragmentation. It insists on the single point of view (at least, in its classical phase) and involves us automatically in a single space. Inasmuch as a three-dimensional space is a concomitant of one dimension in time, we find fragmentation developing in both space and time, and in both poetry and painting. Because of the insistence on single times and single spaces, the possibility of "self-expression" arises. In mannerism, this possibility manifests itself in an insouciant violation of the canons of proportion and color, and a realization of the potential inherent in a variety of visual spaces within a single visual space—fragmentation within set parameters. Mannerism may be construed as a prescient insight which allows for the analytic observation so prevalent in the seventeenth century.

Baroque can be regarded as a countergradient or counterthrust or as a return to balance and classical poise after the onset of visual specialism and mannerist fragmentation. If the entire development of perspective is considered as a steadily developing gradient or environmental stress in the Western world, then the virtuoso effects of the mannerist in literature and in painting alike can be seen as a kind of faltering of that gradient. It is hard for us to realize that perspective in its first stage produced all the shock of novelty and nausea that cubism and abstract art produced in the early twentieth-century period or that perspective now gives in the Arab or Hindu world today. Just how odd and repellent perspective felt is manifested in the passage of *King Lear* in which Edgar seeks to save Gloucester:

EDGAR: Come on, sir; here's the place. Stand still. How fearful
And dizzy 'tis to cast one's eyes so low!
The crows and choughs that wing the midway air
Show scarce so gross as beetles. Halfway down
Hangs one that gathers sampire—dreadful trade!
Methinks he seems no bigger than his head.
The fishermen that walk upon the beach
Appear like mice; and yond tall anchoring bark,
Diminish'd to her cock; her cock, a buoy
Almost too small for sight. The murmuring surge
That on th' unnumb'red idle pebble chafes
Cannot be heard so high. I'll look no more,
Lest my brain turn, and the deficient sight
Topple down headlong.
GLOUCESTER: Set me where you stand.
EDGAR: Give me your hand. You are now within a foot
Of th' extreme verge. For all beneath the moon
Would I not leap upright.

The newly blinded Gloucester, wishing to destroy himself, seeks to leap over the cliff edge. Edgar saves him by deception, describing the hypothetical scene in classical perspective. The fact that a hypothetical scene is being described to a blind man presents an ideal situation for poetic virtuosity. To a blind man, visual space affords an ideal completion or closure of his perceptual modes. Shakespeare sensed the need and relevance of illusion, and he seems to have understood thoroughly that perspective was the art of illusion. But to produce visual illusion by verbal means was a special tour de force for which there is no precedent. The opening words of Edgar to Gloucester: "Here's the place. Stand still," indicate Shakespeare's awareness of the art of perspective.

THE APPLE OF THE EYE AS A BROADCASTING STUDIO

In ancient writings on vision two polar points of view were prevalent. On the one hand, emission theorists regarded the eye itself as the source of rays which explore the world somewhat as the fingers palpate objects. On the other hand, reception theorists regarded the eye as a receiver of information originating from external objects. The classical theory of reception asserted that multiple *eidola* (copies) are detached from objects to approach and finally enter the eyes of observers. In this manner, the eye and the Sensorium (perceiving mind) behind the eye gain knowledge of the object. (Gyorgy Kepes, *Structure in Art and in Science*)

Emission theories prevailed for many centuries. They yielded to reception theories with the advent of Newton's *Optics*. Today revisionists have been making way once more for a serious attitude toward the emission theory.

Prior to such advanced theory, however, the late nineteenth century saw a remarkable advance of Newtonian ideas, with particular emphasis on the afterimage and simultaneous contrast. While this theory is generally known to practicing painters, its wider sociological implications have never been explored. To explain simply, in the field of color the afterimage consists of a physiological balancing on integral white. A brief formula might be sensory impact plus sensory completion equals white ($SI + SC = W$). For example, any hue gets grayer the longer it is observed. This process, as a matter of fact, starts instantaneously. It is postulated that just as white is a result of the assembling of the primary colors in ratio, so touch is an assembly of all the senses in ratio. Black is, therefore, the afterimage of touch. Naturally as the visual gradient of the culture ascends, the modalities of touch are minimized. This appears very vividly in the sensory evolution of the arts. From cave painting to the Romantics, there is steady visual progress. Thereafter, with the coming of synesthesia in the arts and non-visual electronic phenomena in the sciences, we may well be moving into a kind of zero-gradient culture, with all modes of experience receiving simultaneous attention.

The need for physiological and psychological balance means that any *new* sensory impact needs to find *familiar* sensory completion, just as a man on the moon would need to translate all lunar experience into familiar earth terms.

The Lear landscape in Shakespeare offers a close counterpart to the Dürer illustration showing a perspective drawing being made through a transparent screen. The artist fixes himself in position, allowing neither himself nor his model to move. He then proceeds to match dots on the picture plane with corresponding dots on the visual image, a rather bizarre anticipation of the head clamps of Daguerre. This is the kind of "single vision" that William Blake later deprecated as "single vision and Newton's sleep." It consists basically in a process of matching outer and inner representation. That which was faithfully represented or repeated has ever since been held to be the very criterion of rationality and reality. When there is a failure of such correspondence, a person is thought to be either hallucinating or living in a world of self-deception. Yet today even the assembly-line technique of reproducing identical objects is breaking up under the impact of the infinite variability inherent in automated techniques.

MEDIEVAL WORLD IN A REAR-VIEW MIRROR

In their paintings the mannerists used a simple aggregate of traditional classical perspectives, unlike the medievals, whose multileveled images consisted of all-inclusive sensory experiences. The mannerists, therefore, utilized the Medieval parataxis, but in a new visual perspective mode. Mannerism is, as it were, a retrospective of the Medieval world in the new perspective style. It rehearses Medieval art, not as form, but as content of a new artistic order. Such, in its own way, is the work of Spenser, or of almost any major Renaissance artist. Each tends to use the Medieval world, not as his form, but as his content, thus attaining a retrospective of the Middle Ages. In *contrapposto*, one finds an early manifestation of the Baroque, a contrapuntal use of the human body in which the torso is violently twisted to present two postures at once. This is certainly familiar to many people in the poetry of John Donne:

> What if this present
> Were the world's last night?
>
> or
>
> At the round earth's imagin'd corners.

BAROQUE AMBIGUITIES

In *Milton, Mannerism and Baroque,* Roy Daniells remarks on the Baroque use of doubleness, with its architectural and artistic creations in general as a means of taking us into the action suddenly "and in such a way as to make it our own story, yet allowing us to see the whole which is to be unfolded."

Broken or double perspective as a way of involving us in the action of a work of art is, of course, essential to John Donne and the metaphysicals, as in "Whoever comes to shroud me," where the poet assumes that he is simultaneously dead and alive. Milton uses it for the opening of *Paradise Lost,* Daniells observes, as a means of getting us into the center of the poem and at the same time of providing an inclusive vision:

> Of Man's First Disobedience, and the Fruit
> Of that Forbidden Tree, whose mortal taste
> Brought Death into the World, and all our woe,
> With loss of Eden, till one greater Man
> Restore us, and regain the blissful Seat,
> Sing Heav'nly Muse. . . .

A further note on mannerism might help to clarify the meaning of perspective for its time. The mannerists were both enthusiastic and skilled in the new perspective technique of isolating facets and postures of things and events. Their work has proved puzzling to many people, but, in point of fact, the play *King Lear* itself is really a study in mannerist extremes of individualism. Likewise, *Othello* is a study of the breakdown of human order in perception and living that results from an extreme insistence upon one kind of experience. Othello demands from Iago not just proof but "ocular proof" of the infidelity of Desdemona. He finally obtains it, to his own destruction. The insistence on reducing certainty to one sensory mode is almost a prophetic indication of the new science and philosophy that swept into the Renaissance, demanding definitive and quantitative measurement as the criterion of certitude in all things.

THE BAROQUE QUEST FOR OLD HENSYNE

Whereas mannerism looked fleetingly and, as it were, with agitated rapidity at separate moments and facets, Baroque art and poetry sought to unify disparate facets and experiences by directing attention to the moment of change.

The moment of change was a favourite Baroque theme. Bernini, as we have noticed, represents Anchises and Proserpine at the instant of their being carried off, Daphne as the bark folds round her body and her fingers put forth leaves. The action looks both ways and we know from the extreme and subtle expressiveness of Bernini's modelling (anxious old eyes of Anchises, the gripping fingers of Pluto upon Proserpine) both what the subjects have been and what they will be. Milton similarly, and unlike Dante or Spenser, gives us the moment of change—in Satan the moment when the realization of hell bursts upon him, not less than archangel fallen; in Eve when we have been thoroughly prepared to see the moment of eating the fruit as an index pointing to past and future. (Daniells, *Milton, Mannerism and Baroque,* University of Toronto Press, 1963)

Physiologically, the center of the eye is sensitive to texture and color. It is the periphery of the eye that is extraordinarily sensitive to light, dark and movement. In Medieval and primitive times the mode of seeing came from the use of the center of the eye. In the Renaissance and the Baroque periods, the peripheral area achieved primacy. In *Finnegans Wake,* Joyce includes some of these themes in his chapter on Belinda the kindly fowl and her old hensyne. Pastimes are past times.

RISE OF THE STORY BOOK ILLUSTRATORS

The Baroque grandeur and resonance of double perspective and contrapuntal theming were diminished in a reaction toward prettiness of ornament and virtuosity of miniature delicacy. The names of Watteau, Boucher and Fragonard will serve to suggest these modes in painting, as do the Romantics in poetry.

In the midst of this Rococo indulgence and flamboyance there came nostalgia for the severe ideal of single perspective. Stress on the single visual gradient that had preceded the Baroque and the mannerists began to be felt again. With this new stress on the visual gradient, we enter the world that has been called "neo-classical" in both painting and poetry.

Whereas the Baroque sought to capture the moment of change in order to release energy dramatically, neo-classical art sought to eternalize the dramatic moment. In the *Pygmalion* of Falconet, discussed by D. W. Robertson, the artist's ambition is to arrest a moment of vision in order to suggest a train of actions that will follow it. In the Baroque, concern with the moment was intended to involve the audience dramatically. In neo-classical art, the arrest of the moment was intended to involve the audience at a dramatic point in the narrative in order to hold the audience for the narrative that follows. This is the difference between drama and narrative, with neo-classical art tending to *illustrate* fictional narrative. The Pre-Raphaelites also began to concentrate on "the aesthetic moment"—a moment of crux in a developing narrative or a life history. Robert Browning spent his life on this theme.

Where the early Renaissance had discovered the road to outer specialism via fixed points of view, the neo-classical period of art and letters applied a similar method of fragmentation and specialism to the inner feelings and the emotions. William Blake, poet, engraver and painter, fought furiously against the new fragmentation of feeling, abhorring the work and influence of Locke and Newton, and striving to integrate experience in new visions, which in the plastic arts he translated into iconic forms. Blake's foray into the Medieval modes was no archaeological quest or historical recovery. Rather it was the intuitive response of a powerful, artistic sensibility to the challenge of his own times. It is possible that Blake's dual role of poet and painter caused him to be more aware of the need for integral sensory orientation. Certainly his insistence on the "bounding outline" is a statement of his belief in the tangible nature of reality, which had become insubstantial in the hands of the late Baroque painters. To him, Rembrandt was anathema because he lost the defining edge of forms in the seductive world of shadows. Such a return to the iconic or reified symbolic forms was also an anticipatory thrust into the twentieth century.

The discovery of the fixed position for visual experience, which is perspective, forces peripheral vision upon the viewer. The corollary of this was a growing awareness of the peripheral area for active exploration, both psychically and physically. This dualism between center and margin found its parallel in the break between the subject and the object. It is only perspective which allows for dispassionate survey and noninvolvement in the world of experience. In Skira's *Seventeenth Century* we find "dualism between outer forms of life and the problems of the soul, which now were every man's personal affair." Politically and economically, the center-margin dynamic of the spongelike action became constitutive in eighteenth-century organization.

AS IF EVERY MOMENT WERE HIS LAST

In eighteenth-century painting, the time depicted was typically narrow and the thrust of action of the figures suggested imminent motion. The Chinese say Westerners are always getting ready to live—"Living as if every moment were his last" (*Finnegans Wake*). By contrast, in preliterate art forms, the time depicted includes all the possible moments of that thing's existence, rendered with iconic outline. The action of figures does not suggest any future motion but rather depicts the significant profile of an action that is timeless, and therefore inclusive of all possible times and all possible actions in all possible spaces.

The Romantic poets sought to isolate single emotions and therefore found it necessary to use perspective. When the content of such work is taken from the slum environment, the result is called "sensationalism." To bring any part of the environment into a privileged area, whether on the printed page or the page of canvas, is to create the effect of reportage. This is tantamount to saying the newspaper is itself a Romantic art form. Goya's *Horrors of War* are essentially reportage, even by our journalistic standards.

The neo-classical in painting often meant the use of classical themes as content for Romantic forms—for example, in David's painting, *The Oath of the Horatii*. In the same way, the Pre-Raphaelites adapted Medieval content for Romantic forms of treatment. And the Romantic poets, repelled by the new industrial environment, seized as their content the preceding agrarian environment of nature and the handicrafts.

One of the interesting features of art history is the antipathy toward the Baroque forms in some Northern areas of Europe, including Germany. The Counter Reformation in the seventeenth century was not entirely separate from the countergradient of Baroque art. The double perspective and dynamic interplay of the Baroque are absent from the neo-classical revival of the eighteenth century. If the Romantics concentrated on one feeling or emotion at a time in their poetry and painting, historians like Gibbon managed to seem classical by using one perspective at a time. Like the neo-classical painters, Gibbon used single perspective and ancient content for his art. In fact, it is difficult to conceive how the Romantics could have achieved their concern with the distant in time and space except by means of single perspectives. It is this that makes Byron and Delacroix and Gibbon and Scott and Géricault closely related figures.

STOPPING THE WORLD IN THE AESTHETIC MOMENT

Newton's *Optics* had an extraordinary influence on eighteenth- and nineteenth-century poetry and painting alike. His revelation of the natural power of the eye to refract the visual world encouraged artists to select outer landscapes that isolated a particular mood or feeling from the emotional spectrum. Whereas the Baroque artists had made the great discovery about the moment of change, the eighteenth century became excited about the aesthetic moment, the moment of arrested awareness, as a moment of art emotion. Refraction was extended at once to the mental and emotional life of man. The external world was studied for its powers to select and to refract particular qualities of experience. Wordsworth's note on Lucy, "A violet by a mossy stone/ Half hidden from the eye," is a striking example of the isolation of delicate qualities by the juxtaposition of natural objects. Marjorie Nicolson's book *Newton Demands the Muse* provides a fascinating account of the poetic response to Newton's *Optics*. Insofar as Newton encouraged a more intense stress on visual and uniform space, he contributed to what I. A. Richards calls the "neutralization of nature," which is manifested in the Impressionists of the late nineteenth century. It was the isolation of particular moments and qualities of experience that helped to build up the eighteenth-century interest in aesthetic stasis and detachment from the world. In nineteenth-century art, emotion tended to be associated more and more with unworldly and otherworldly attitudes. In his *World as Will and Idea* Schopenhauer made the aesthetic moment of stasis memorable as the prime means by which one could, as it were, stop the world and get off.

A VIOLET BY A MOSSY STONE
ROMANTIC INTEREST IN ART PROCESSES AS IN THE GROWING-UP PROCESSES

It is almost as a reaction to the stress on the aesthetic moment that the Romantics developed a deep concern with the creative process in both art and life. It is especially familiar in J. J. Rousseau's *Émile*. Newton had strengthened interest in environmental conditions of growth and nurture. Blake's *Songs of Innocence* and *Songs of Experience* are familiar as part of this new interest in the growing child and the effects of established institutions on human potential: "Shades of the prison-house close round the growing boy." The painter Chardin manifests interest in the effects of home, work and play such as appear in Robert Burns' "The Cotter's Saturday Night." Concern with the processes in the arts led some nineteenth-century aestheticians to consider that while the genius had a mind that used associations, his chain of associations worked backwards. Their concern was enriched by the work of Samuel Taylor Coleridge. The studies of Coleridge inspired Edgar Allan Poe. The effect of Poe on Baudelaire and Valéry is well known.

The idea of a work of art as a direct manifestation of the creative process itself exerted wide influence among the Symbolists. It remained only to devise means to include the audience in this creative process in order to reach that stage of aesthetics that is familiar in Expressionism and in the speculations of the twentieth century.

CONCERN WITH A NARROW SLICE OF TIME

Delacroix's attraction to the *plein air* sketches of Constable tended to inject the later immediacy of the Impressionist mode into the Romantic. Constable, by paying great attention to the effect of light on natural colors, moved painting away from a Chardin-like concern with the object per se. A parallel to Constable can be found in Turner, where the subject matter has been almost totally eliminated in favor of the moving lines of force of the general environment: storm, steam, mist, water, light, etc., play a primary role in the paintings of Turner.

Whistler is not unrelated to Turner—he too shows predilection for the lines of force that shape the outer scene, albeit with a more delicate taste. Preoccupation with lines of force when filtered through a very refined sensibility, such as Whistler's, naturally attracted the painter to Oriental models with their concomitant stress upon formal, structural factors. As the interest in formal factors mounted, the interest in subject matter declined to the point where Whistler could name the portrait of his mother *Arrangement in Black and Gray*.

SEURAT AND THE RISE OF FORMAL SPACE, OR THE END OF NEUTRAL NEWTONIAN SPACE

Seurat, by a completely different route, arrived at a similar point of stress. By utilizing the Newtonian analysis of the fragmentation of light, he came to the technique of divisionism, whereby each dot of paint becomes the equivalent of an actual light source, a sun, as it were. This device reversed the traditional perspective by making the viewer the vanishing point. C. S. Lewis in *The Discarded Image* makes a similar observation when contrasting modern man with Medieval man. He explains that the model of space gradually created by Medieval man gave one the feeling of *looking in*. In contrast, he points out that modern man feels that he is *looking out*: "Like one looking out from the saloon entrance on to the dark Atlantic, or from the lighted porch upon the dark and lonely moors." This reversal of spatial perspective also occurs in the poet Hopkins, whose favorite term "inscape" (as compared with "landscape") draws attention to the same change of perspective as that created by Seurat. Seurat returned to the paratactic Egyptian image. By pushing visual modes to their extreme, Seurat returned to the most ancient forms of space and time and figure. The Impressionists painted light *on*. Seurat painted light *through*, making paint itself the light source, and anticipating Rouault's recovery of the stained-glass-window effect of light through. Now the painting looks at the audience once more, in contrast to the seventeenth-century portrait in which the sitter, but not the painting, looked at the audience. We are suddenly in the world of the "Anxious Object" which is prepared to take the audience inside the painting process itself. The Expressionist's aim of including the audience in the process of making art has seemed to many people to bring the same degree of distasteful ugliness into the art world as psychiatry brought into the psychological world. It is, perhaps, an occasion on which to remind ourselves that perspective space seemed forbidding and unfriendly to Gloucester and Edgar in *King Lear* and to the lover in "To His Coy Mistress."

Beginning with Seurat, then, space is no longer neutral. We are inclined today to regard paintings as radiant forms of energy much in the way that the Oriental world does. Perhaps the most obvious example of how space has ceased to be neutral, in the old visual and Newtonian sense, is to be found in the world of the astronaut. The totally designed environment necessary to life in the space capsule draws attention to the fact that the astronaut makes the spaces that he needs and encounters. Beyond the environment of this planet there is no space in our planetary or "container" sense. The gravitational point once transcended, the astronaut must have his own environment with him, as it were. There is no upside down in Eskimo art or in a space capsule any more than there is perspective or foreshortening (weight, or gravitational force, came in with perspective). Strong indications are given to the astronauts that objects, as well as people, create their own spaces. Outer space is not a frame any more than it is visualizable.

CHINESE WEDDING OF THE MEDIEVAL DOLL

Artists and painters had made many explorations in nonvisual space by the end of the nineteenth century. Many of these studies center in Impressionism and in Oriental influence in style and composition. For example, the hokku form has been called the form of "superpositions." It is a paratactic way of arranging objects and situations, colors and rhythms, without counteraction. It is based on the discovery that things radiate energies and create their own spaces even without the benefit of visual or lineal ties. As Earl Miner says in *The Japanese Tradition in British and American Literature* (Princeton University Press, 1958), "Pound used the super-pository method, as it may be called, as a very flexible technique which provides the basic structure for many passages and many poems." Discussing this kind of structure in Eliot's *Waste Land,* I. A. Richards had referred to it as "the music of ideas." It seems to be much easier for literate people to grasp the nonvisual in art if it is mediated through the world of sound. In effect, the superpository method heralded the advent of the awareness of electronic structures that exceed the scope of our visual powers. In literary terms, the revival of parataxis means the decline of the "story line" as a means of organizing verbal structures. If it is understood as the decline of visual preference in favor of the participation of the other senses in art, it is easier to comprehend how Chinese and Japanese art (along with medievalism) contributed to the later nineteenth-century cult of Aestheticism. Earl Miner observes:

> ... This was also the age of Whistler and the Japanese, the time when the advance-guard called itself by such names as Aestheticism and Impressionism. Aestheticism seems to have been composed of two parts of medievalism and one of Orientalism, a combination particularly vulnerable to the charge of preciosity which was brought against it.

It is easy to see why an art that diminished the role of the visual in favor of the tactile got tagged with the term "Fleshly School." The Chinese vogue had been felt during the eighteenth century. The vogue of *Japanisme* grew steadily from 1865 to 1895 and became associated with many of the dominant names in painting and poetry. Duret declared that "the Japanese are the first and finest Impressionists," and Van Gogh felt that Japanese art was the "true religion." In Japan, during the first period of strong Western influence, the block print fell into the contempt that our painters have often felt for the photograph. That is to say, as the Japanese increased the visual stress of their culture, they became alienated from their own sensory tradition. By the same token, as the Western world entered the domain of electronics, it began to feel alienated from much of its own visual tradition.

James McNeill Whistler was a key figure in popularizing Japanese art in England. Sadakichi Hartmann said of Whistler's major work: "Never have the elements of Eastern and Western art been so originally united as in these poems of night and space."

WILDE'S JAPANESE BUTTERFLY

It was Whistler who introduced Swinburne and Oscar Wilde to Japanese modes of perception. "Fly away, butterfly, back to Japan," Swinburne later wrote to James McNeill Whistler. The technique of the Imagists owes much to these Oriental currents, and Wyndham Lewis wrote: "Pound's nearest American analogue in the past is not Whitman, however, or Mark Twain, but a painter, James McNeill Whistler.... Like Pound in the literary art, it was in the extreme-orient that Whistler discovered the fundamental adjustments of his preference." And Eliot, of course, beginning with Omar, had learned a great deal from the same Oriental means of manipulating images. Pound and Eliot were in closest association just before and during the First World War. In 1914 Pound wrote in *The Egoist:*

> I trust that the gentle reader is accustomed to take pleasure in "Whistler and the Japanese." Otherwise he had better stop reading my article until he has treated himself to some further draughts of education.
> From Whistler and the Japanese, or Chinese, the "world", that is to say, the fragment of the English-speaking world which spreads

itself into print, learned to enjoy "arrangements" of colours and masses.

ABSTRACT ART AS THE INTERNALIZATION OF VISUAL SPACE

Visual space is being internalized at the terminus of the visual gradient. This reverses the development of perspective, which in its time was first internalized by Western cultures as they developed their visual gradient. What had been experienced by populations was next externalized by artists. This may, indeed, be a primary function of art: to make tangible and to subject to scrutiny the nameless psychic dimensions of new experience. Pictorial space couldn't be "seen" at first. Initially it existed as something objective, alien and uninviting. In his *Saving the Appearances,* Owen Barfield contrasts the "participation mystique" of preliterate populations with the objective detachment that comes with phonetic literacy. After centuries of participation in the cosmic energies, the rise of civilization reshaped human awareness visually.

GALILEO'S HEAVENLY JUNKYARD

Today, with the decline of the role of visual power in an age of electronic circuitry and participation in many nonvisual dimensions of space and time, Western populations are once more inclined toward involvement in cosmic energies in their arts and entertainment. However, it was science and engineering that created the new experiential environments of work and living that are now being increasingly reflected in the arts. In a prophetic way this new involvement of whole populations in these hidden dimensions of energies had been fully indicated by artists and poets in the later nineteenth century. Whereas in the Renaissance it was the encounter with the new pictorial or visual space that created discomfort and dismay, the reverse is true in our time. It is the rediscovery of nonvisual, multisensuous spaces that bothers and confuses us. Kohler makes an interesting observation (in *Gestalt Psy-*

chology, Mentor Books, 1947) on the new space as experienced in the seventeenth century

> ... What was so shocking in Galileo's astronomical discoveries? That he found so much going on in the sky, and that as a consequence the astronomical order was so much less rigid than people had been able happily to believe before. If the heavens began to show such a lack of rigid reliability, if they approximated the restlessness of terrestrial conditions, who could feel secure in his most important beliefs? Thus primitive fear inspired the furious attacks which the Aristotelians of his time directed against Galileo. It seems quite possible that the excitement produced by Harvey's discovery of the circulation of the blood contained a similar element of fear, because this discovery suddenly disturbed the conception of man as a rigid structure. With so much unrest in his interior, did not life as a whole become a most precarious affair?

If the telescope of Galileo had revealed the heavens as a kind of junkyard, in lieu of the crystalline spheres, there was also a dismay because of the absence of any sublime astral music. Instead of the heavenly symphony, the heavens were revealed as an area of ghastly silence: "The eternal silence of these infinite spaces terrifies me," says Pascal. Both visual and auditory order seemed to be denied by the new astronomy.

THE PUBLIC AS DITHYRAMBIC SPECTATOR

When the visual gradient achieved a considerable stress in sixteenth-century culture, an increasing distance developed between art and the public. As the public became patron of the arts via the marketplace, this gap between public and art widened. Fine art became reserved for elites and connoisseurs. Today one extreme development has been Pop Art, which accepts the utmost banality and vulgarity of the environment as a quasi-art form. This kind of art draws attention to an electronic world in which the programing of whole environments by information is as feasible as the earlier programing of classrooms. Art is ceasing to be a special kind of object to be inserted in a special kind of space. The sense of participation in the art process has reached an extreme in the so-called "Happenings," which are plausible simulations of environmental control. But in contrast to Jackson Pollock, Pop Art may have misled many people trying to orient themselves in the new spaces of our world. Pop Art tends to organize the new concept of space generated by our exploration of outer space by filling it with tangible artifacts. Unless God's "Copernican junkyard" has a parallel to "God's Little Acre," which was constantly being transplanted, such interpretation can lead only to confusion.

Action painting seems to strive to externalize the psychic process of making as something to be contemplated in place of the old world of objects. Another means, parallel to the Happening, of putting the audience into action as part of the art process is the Theater of the Absurd. As in the work of Kafka or Hieronymus Bosch, the Theater of the Absurd brings contrasting environments of space and culture into direct encounter. Whole cultures become characters, as it were, in a play. The result is that no one makes sense to anyone else. The characters merely sound off simultaneously like the unrelated news stories in the daily paper. Their only coherence is in the dimension of the date line, or in their simultaneity. Today spaces and environments are put up with the same freedom and speed as sets on movie lots. The human response to this environmental flux is the sense of the absurd. The Happening is the absurd in action. Some practitioners insist that the organization of a Happening requires the precision of launching a man into space. They could also add that the one is no more absurd than the other.

Man in outer space as yet has no means of *imagining* the nature of his own experience in space. Until artists have provided him with adequate forms to express what he feels in space, he will not know the meaning of the experience. The work of Jackson Pollock may, in this regard, have some relation to the music of John Cage. Both, in their works, may be groping for an objective formal means of defining new dimensions of experience that have no rapport whatever with previous spaces or modalities of sense perception. The artist today might well inquire whether he has time to make a space to meet the spaces that he will meet. "Paul Klee declares, 'I want to be as the newborn, knowing nothing, absolutely nothing about Europe . . . to be almost primitive.' " (Harold Rosenberg, *The Tradition of the New*)

TOWARD A SPATIAL DIALOGUE

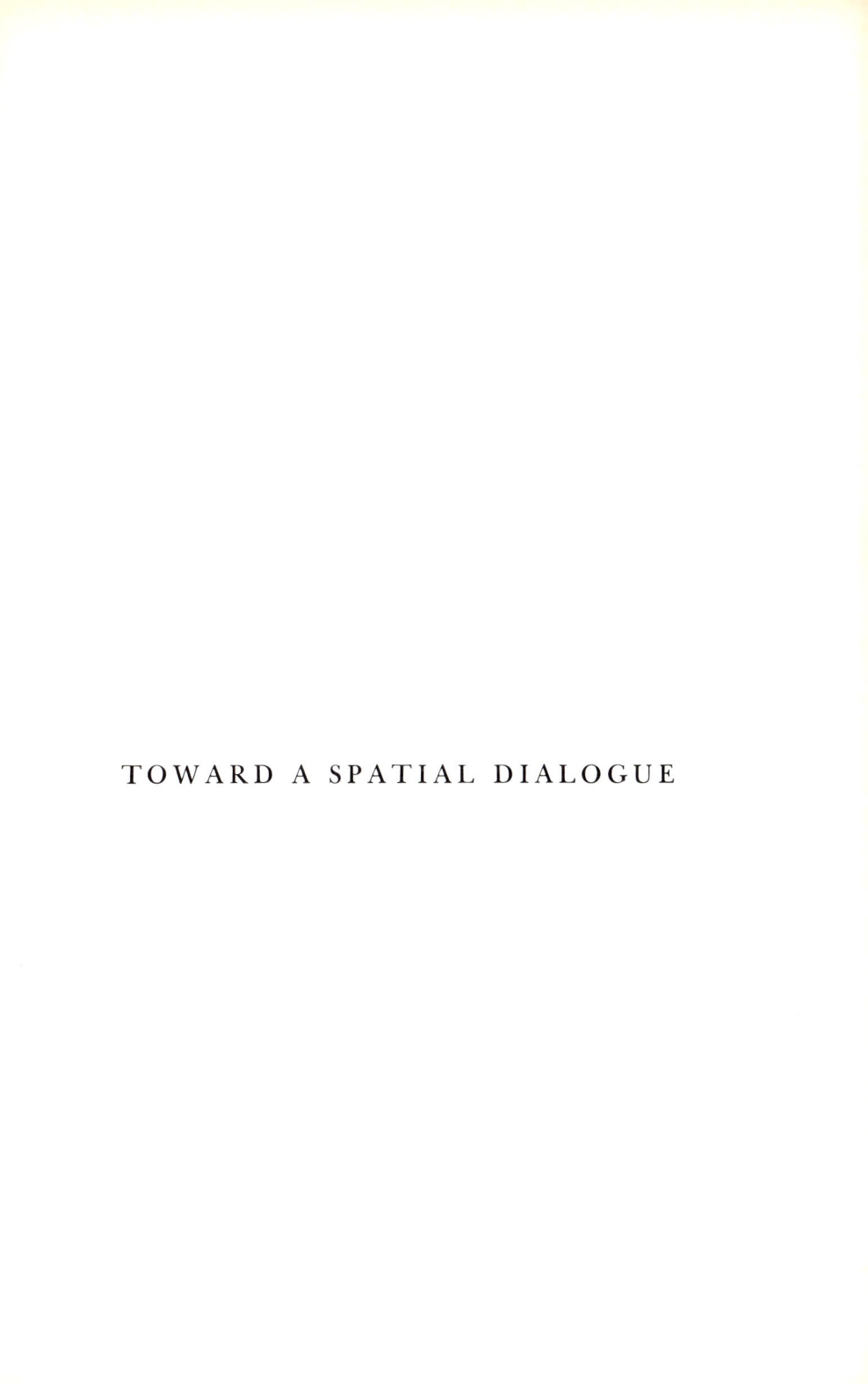

1 Beast of the Sea,
Come and offer yourself in the dear early morning!
Beast of the Plain!
Come and offer yourself in the dear early morning!

Great Sea
Sends me drifting,
Moves me,
Weed in a river am I.

Great nature
Sends me drifting,
Moves me,
Moves my inward parts with joy.

The lands around my dwelling
Are more beautiful
From the day
When it is given me to see
Faces I have never seen before.
All is more beautiful,
All is more beautiful,
And life is thankfulness.
These guests of mine
Make my house grand.

The word itself as evocative power, not a sign. **1**

Fusion with the natural process. "Weed in a river am I."
An artist might have said, "Used by the words am I."

No question here of privacy or private identity, but a free flow of corporate energy. Here the people make their world and are not contained *in* it. The cosmos becomes an extension of their energies.

"A poem should not mean
But be."
—ARCHIBALD MACLEISH

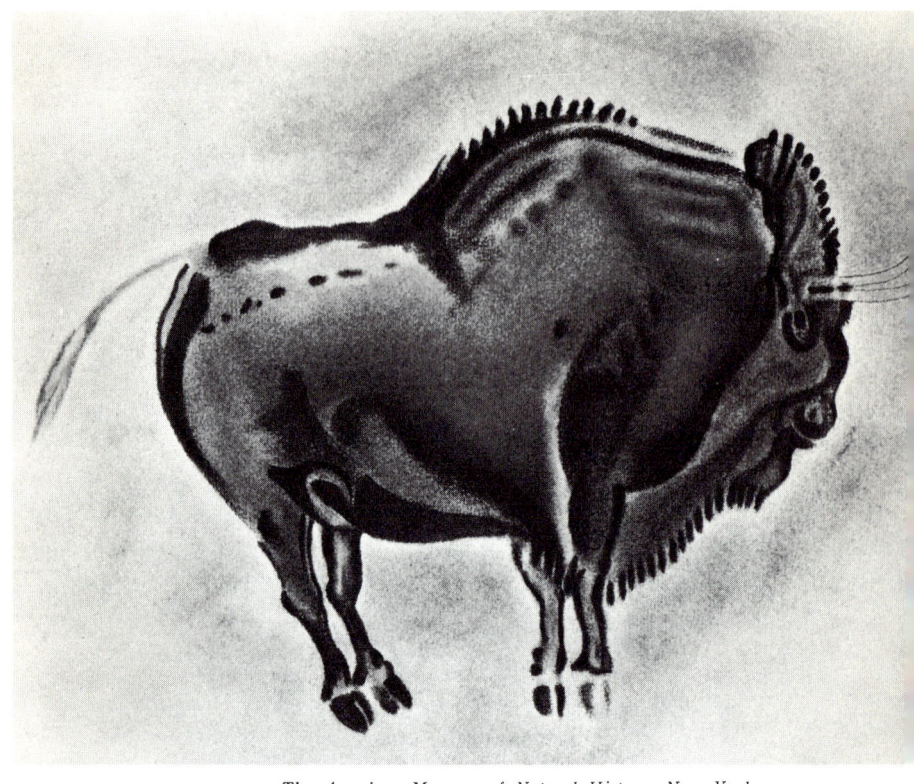

BISON, ALTAMIRA CAVES, SPAIN. *The American Museum of Natural History, New York*

1

Eskimo says, "Weed in a river am I." He could also say, "Used by my kayak am I."

The magic of the cave image lies in its *being*, not in its being seen.

The hand has no point of view.

The cave painters used somewhat specialist technology. Were they on the borders of Neolithic time?

2 Like the higher apes, the alphabetic cultures knew how to let go.
allforabit=alphabet

Consider the reversal of this cultural gradient by the poetry that goes *allforabyte*.

"*Mythos*" is a Greek word for "word." All words originally revealed complex processes and were called "momentary deities" or epiphanies.

> "The apparition of these faces in the crowd;
> Petals on a wet, black bough."
> —EZRA POUND

Earl Miner quotes Pound's account of the origin of this "hokku-like sentence":

> Three years ago [1911] in Paris I got out of a "metro" train at La Concorde, and saw suddenly a beautiful face and another and another . . . and I tried all day for words for what that had meant for me. . . . And that evening . . . I found suddenly the expression . . . not in speech but in sudden splotches of colour. It was just that—a "pattern" or hardly a pattern if by pattern you mean something with a repeat in it. But it was a word, the beginning for me of a new language in colour.

Pound observes further:

> The Japanese have the same sense of exploration. They have understood the beauty of this kind of knowing [i.e., "imagistic" as opposed to "lyric" writing]. . . . The Japanese have evolved the form of the hokku.
>> The fallen blossom flies back to its branch:
>> A butterfly.
>
> That is the substance of a very well-known hokku. . . .
> The one image poem is a form of super-position; that is to say it is one idea set on top of another. I found it useful for getting out of the impasse left by my metro emotion.

CHINESE CHARACTER

2

If the Altamira Buffalo is a prayer, the ideograph is a
probe and the later alphabet is a package.

With the ideograph we begin to move from the
reverential to the referential.

As the bounding line ceases to be stressed, there is
steady lessening of involvement in process.
There is steady strengthening of visual values
and detachment.

In contrast to phonetic letters,
the ideograph is a vortex that responds to lines of force.
It is a mask of corporate energy.

ODE ON A GRECIAN URN
John Keats

Thou still unravish'd bride of quietness,
   Thou foster-child of silence and slow time,
Sylvan historian, who canst thus express
   A flowery tale more sweetly than our rhyme:
What leaf-fring'd legend haunts about thy shape
   Of deities or mortals, or of both,
      In Tempe or the dales of Arcady?
   What men or gods are these? What maidens loth?
What mad pursuit? What struggle to escape?
   What pipes and timbrels? What wild ecstasy?

Heard melodies are sweet, but those unheard
   Are sweeter; therefore, ye soft pipes, play on;
Not to the sensual ear, but, more endear'd,
   Pipe to the spirit ditties of no tone:
Fair youth, beneath the trees, thou canst not leave
   Thy song, nor ever can those trees be bare;
      Bold lover, never, never canst thou kiss,
Though winning near the goal!—yet, do not grieve;
      She cannot fade, though thou hast not thy bliss,
         For ever wilt thou love, and she be fair!

When asked what music he liked best, Mozart replied: "No music." **3**

"Heard melodies are sweet, but those unheard are sweeter" reaches the sense of touch through the word of the ear. To the sense of touch the significant form is found in the interval, not in the connection.

Keats was looking eagerly for a story line, where the vase probably provided only an incident. A coincidence elevated to visual dignity acquires the character of a story line.

Finally, the vase-makers painted a series of pots to create a plot. (**See 4.**)

THE KILLING OF THE WOOERS (SCYPHOS VASE), LEFT SIDE. *Staatliche Museen, Berlin*

THE KILLING OF THE WOOERS (SCYPHOS VASE), RIGHT SIDE. *Staatliche Museen, Berlin*

3

There is no point of view from which to look at a vase.

The viewer can begin or end anywhere.

The three-dimensional Greek figure stands stark on a single plane.

The vase moves toward Euclidean space, but "Euclid alone hath looked on beauty bare."
—EDNA ST. VINCENT MILLAY

This art moves steadily toward *matching* or representation, but retains a great residue of primitive *making* and multisensuousness.

BYZANTIUM

William Butler Yeats

The unpurged images of day recede;
The Emperor's drunken soldiery are abed;
Night resonance recedes, night-walkers' song
After great cathedral gong;
A starlit or a moonlit dome disdains
All that man is,
All mere complexities,
The fury and the mire of human veins.

4

"All art that is not mere story-telling, or mere portraiture, is symbolic . . . for it entangles, in complex colours and forms, a part of the Divine Essence."
—W. B. YEATS, "Symbolism in Painting"

"Sailing to Byzantium" in the twentieth century has been a purging of the visual images of the Western day. The splendor of the Eastern resonance has been restored to the arts of the Western world in poetic rhythms and symbolic forms.

"Images of day," "night resonance"—as one sense dims down, another moves in to fill its place.

Gandhara embraced Western visual connectedness in the first century A.D. Yeats rejected it in the twentieth century in favor of the icon and the interval: "I have spent my life in clearing out of poetry every phrase written for the eye. . . ."
—"An Introduction to My Plays"

SCENE FROM THE LIFE OF BUDDHA (RELIEF FRAGMENT). *Royal Ontario Museum, University of Toronto*

The Roman meeting with the East led to an Eastern emphasis on the interval that is mated to a Western concern with psychological connectives.

4

The Gandharan sculptural story line finds its echo in the façades of Gothic churches.

Compared to the explicit statement of the Greek vase, the story line is implied in the psychological gesture, or posture.

5

Moreover when ye fast, be not, as the hypocrites, of a sad countenance: for they disfigure their faces, that they may appear unto men to fast. Verily I say unto you, They have their reward.
—ST. MATTHEW vi:16

The verb *exterminare* has a quite different meaning from that in which it is commonly understood. Exiles are exterminated, when they are sent beyond the terms or borders of their country. So where *exterminare* is used in the writings of the Church, we ought always to substitute *demoliri,* to destroy, this being the equivalent of the Greek *aphanizein.* The hypocrite "destroys" his face in order to simulate sadness, and carries grief in his countenance when perhaps his heart is rejoicing.
—SOLOMON GOLDMAN, *The Golden Chain*

... In the Good Friday sermons one notices his use of the *Liber Charitatis* theme, which was a traditional one, whose origin is obscure but which had been handled with great effect in a Good Friday sermon by John Fisher, who had likened the crucified body of Christ to a book, and the wounds to capital letters in red. Andrewes, taking up this theme says: "... the print of the nails in them, are as capital letters to record His love towards us. For Christ pierced on the cross is *liber charitatis,* 'the very book of love' laid open before us." Again, "... being spread and laid wide open on the Cross, He is *Liber charitatis,* wherein he that runneth may read. ... Every stripe as a letter, every nail as a capital letter. His *livores* as black letters, his bleeding wounds as so many rubrics."
—PAUL A. WELSBY, *Lancelot Andrewes—1555-1626*

Bishop Andrewes does a multileveled exegesis while wittily embodying the whole in the idea of the format of the book itself.

5

The Medieval scribe, like the ancient scribe, accepted a multileveled approach to his text. Where we find a simple statement, they discovered implications.

The sophistication of the scribe exceeded that of the modern philologist.
See D. W. Robertson, *Preface to Chaucer*.

BRONZE CRUCIFIXION PLAQUE, ST. JOHNS, COUNTY ROSCOMMON. *National Museum of Ireland, Dublin*

ST. MARK, FROM THE GOSPEL BOOK OF THE ARCHBISHOP OF EBBO OF REIMS. *Municipal Library, Épernay, France*

CRUCIFIXION

Repetition of pattern has all the kinesthetic impact of the walking human, all the tactility of the measuring hand span.

The contour tends to be inclusive and timeless.

"The function in Medieval art is to involve all the senses in order to convince."
—GEORGES POULET

ST. MARK

This is an example of the effect of Roman visual on iconic form.

A three-dimensional form in a three-dimensional spatial illusion. There is even a distant landscape background.

This painting reflects the rational lucidity and spatial connectedness of Roman literate culture and imperial bureaucracy.

The icon of the *Crucifixion* is an intensive manifold in contrast to the exploding energy of the pictorial form in *St. Mark*.

Aquinas was later to explain that the literal *(littera)* sense contained all levels of meaning.

6 RUBÁIYÁT OF OMAR KHAYYÁM
Edward FitzGerald

XII
A Book of Verses underneath the Bough,
A Jug of Wine, a Loaf of Bread—and Thou
   Beside me singing in the Wilderness—
Oh, Wilderness were Paradise enow!

XIII
Some for the Glories of This World; and some
Sigh for the Prophet's Paradise to come;
   Ah, take the Cash, and let the Credit go,
Nor heed the rumble of a distant Drum!

XIV
Look to the blowing Rose about us—"Lo,
Laughing," she says, "into the world I blow,
   At once the silken tassel of my Purse
Tear, and its Treasure on the Garden throw."

XV
And those who husbanded the Golden Grain,
And those who flung it to the winds like Rain,
   Alike to no such aureate Earth are turn'd
As, buried once, Men want dug up again.

6

Like the hokku, its quatrains omit connectives and make a series of symbolic leaps.

The youthful Eliot found Omar Khayyám an "almost overwhelming introduction to a new world of feeling." Omar hoicked him out of the merely visual spaces of Western poetry.

In Omar, as in Eliot, there are scarcely any visually connected spaces.

Many Victorians sailed to Byzantium with Omar Khayyám before Yeats set up his tour.

TWO WARRIORS FIGHTING IN A LANDSCAPE (PERSIAN MANUSCRIPT). *British Museum, London*

6

The felt profiles of the flat image are multisensuous; i.e., tactility includes all the senses as white light incorporates all colors.

In Western European art, three-dimensional illusion of depth comes by abstracting the merely visual from the other senses.

Von Békésy, the acoustic scientist, illustrates multidimensional auditory space with two-dimensional Persian painting.

If the three-dimensional illusion of depth has proved to be a cul-de-sac of one time and one space, the two-dimensional features many spaces in multileveled time.

7 THE CANTERBURY TALES
Geoffrey Chaucer

Bifel that in that seson on a day,
In Southwerk at the Tabard as I lay
Redy to wenden on my pilgrymage
To Caunterbury with ful devout corage,
At nyght was come in-to that hostelyre
Wel nyne and twenty in a compaignye,
Of sondry folk, by aventure yfalle
In felaweshipe, and pilgrimes were they alle,
That toward Caunterbury wolden ryde;
The chambres and the stables weren wyde,
And wel we weren esed atte beste.
And shortly, whan the sonne was to reste,
So hadde I spoken with hem everichon,
That I was of hir felaweshipe anon,
And made forward erly for to ryse,
To take oure wey ther as I yow devyse.
  But nathelees, whil I have tyme and space,
Er that I ferther in this tale pace,
Me thynketh it acordaunt to resoun,
To telle yow al the condicioun
Of ech of hem, so as it semed me,
And whiche they weren, and of what degree,
And eek in what array that they were inne. . . .

7

Chaucer's narrative voice imposes the illusion that he is sharing a single space with the reader. His dramatic changes of tone and mask occasionally dispel this illusion and confuse the twentieth-century reader.

The Pilgrims are Pilgrims of eternity, actually out of time. At the inn, or on the road, they *make* the human city. Like Skelton, or Dickens, Chaucer is a crowd.

Duccio de Buoninsegna. MARY RECEIVING THE ANNOUNCEMENT OF HER DEATH. *Opera del Duomo, Siena*

The reverse perspective of the bench, an habitual mode of Oriental art, locates the vanishing point in the viewer. (Seurat, in the nineteenth century, attained this end by quite different means. **See 37.**)

Duccio's discovery of how to place these figures in an architecturally enclosed space moved toward theatricality and the proscenium-arch space in painting.

The sense of the downward thrust of weight generated by perspective strengthens the simulation of a human condition.

The portrayal of proprioceptive tension and body percept leads to an emphatic encounter with the weight of the figures.

BALADE DE BON CONSEYL

Geoffrey Chaucer

Flee fro the prees, and dwelle with sothfastnesse;
Suffyce unto thy good, though it be smal;
For hord hath hate, and climbing tikelnesse,
Prees hath envye, and wele blent overal;
Savour no more than thee bihove shal;
Reule wel thyself, that other folk canst rede;
And trouthe thee shal delivere, it is no drede.

Tempest thee noght al croked to redresse,
In trust of hir that turneth as a bal:
Gret reste stant in litel besinesse;
Be war also to sporne ayeyns an al;
Stryve not, as doth the crokke with the wal.
Daunte thyself, that dauntest otheres dede;
And trouthe thee shal delivere, it is no drede.

The verbal counsels are as iconic as the images. There is no self-assertion, no private lyric stress. All is liturgical and corporate.   **8**

Until author becomes conscious of his personal appearance there is no "self" expression.

CHRIST BEARING THE CROSS (SPANISH SCHOOL, FOURTEENTH CENTURY). *The National Gallery of Canada, Ottawa*

The nonperspective, i.e., nonpictorial, space and
time insure the weightlessness
of the cross.

8

Each face and costume is a separate spatial entity
without individualistic stress.

A realistic expression of anguish would create
a mannerism of time and space.

The rhythmic undulations of the halos is
the visual equivalent of the auditory mumble.
This is the world of the ear and the crowd.

SONG FROM LOVE'S LABOUR'S LOST
William Shakespeare

When icicles hang by the wall,
   And Dick the shepherd blows his nail.
And Tom bears logs into the hall,
   And milk comes frozen home in pail,
When blood is nipp'd and ways be foul,
Then nightly sings the staring owl:
      'Tu-who!
Tu-whit, tu-who!' a merry note,
While greasy Joan doth keel the pot.

When all aloud the wind doth blow,
   And coughing drowns the parson's saw,
And birds sit brooding in the snow,
   And Marian's nose looks red and raw;
When roasted crabs hiss in the bowl,
Then nightly sings the staring owl:
      'Tu-who!
Tu-whit, tu-who!' a merry note,
While greasy Joan doth keel the pot.

9

There is as little story line as there is perspective of space or time.

The multiforms, along with the repetitions, make the song a "Happening."

The mode of song and of festival is inclusive, all-at-onceness. As a means of creating involvement and participation, nothing seems to rival a simple catalogue. Compare "The Seven Ages of Man" in *As You Like It*.

The Roman Martial did it the same way:

"Milk from the flawless firstling of the herd.
Honey, the amber soul of perfumed meads,
And water sparkling from its maiden source . . ."

How much specialism of space and time occurs in the phrase "When icicles hang by the wall"?

Pol Limbourg. FEBRUARY, FROM THE VERY RICH BOOK OF HOURS OF THE DUKE OF BERRY.
Condé Museum, Chantilly, France

9

The cumulative effect of a seasonal or festive experience is achieved by inventory or unconnected assembly of components.

Traditional iconic modes are framed in the beginnings of the new Renaissance pictorial space.

Is Limbourg illustrative or evocative of space and time? Is Shakespeare more or less so than Limbourg?

First appearance of the snow landscape.

10 DANTE'S PURGATORIO

CANTO XII

Even in step, like oxen which go in the yoke,
   I went beside that burdened spirit, so long as
   the sweet pedagogue suffered it.

But when he said: "Leave him, and press on, for
   here 'tis well that with sail and with oars, each
   one urge his bark along with all his might";

erect, even as is required for walking I made me
   again with my body, albeit my thoughts remained bowed down and shrunken.

I had moved me, and willingly was following
   my master's steps, and both of us already
   were showing how light of foot we were,

when he said to me: "Turn thine eyes downward: good will it be, for solace of thy way,
   to see the bed of the soles of thy feet."

As in order that there be memory of them, the
   tombs on the ground over the buried bear
   figured what they were before;

wherefore there, many a time men weep for
   them, because of the prick of remembrance
   which only to the pitiful gives spur;

so saw I sculptured there, but of better similitude
   according to the craftsmanship, all that which
   for road projects from the mount.

I saw him who was created nobler far than
   other creature, on one side descending like
   lightning from heaven.

I saw Briareus, transfixed by the celestial bolt,
   on the other side, lying on the earth heavy
   with the death chill.

I saw Thymbraeus; I saw Pallas and Mars,
   armed yet, around their father, gazing on the
   scattered limbs of the giants.

I saw Nimrod at the foot of his great labour,
   as though bewildered, and looking at the
   people who were proud with him in Shinar.

O Niobe, with what sorrowing eyes I saw thee
   graven upon the road between seven and
   seven thy children slain!

ARGUMENT

Philip H. Wicksteed

Dante has bent down in a sympathetic attitude of humility to converse with Oderisi, and when Virgil bids him make better speed he straightens his person so far as needful to comply, but still remains bowed down in heart, shorn of his presumptuous thoughts (1-9). As he steps forward with a good will, Virgil bids him once more look down at the pavement which he is treading, and there he sees as it were the lineaments of the defeated proud, from Lucifer and Briareus to Cyrus and Holofernes and Troy. The proud are laid low upon the pavement as the humble were exalted to the upspringing mountainside (10-72). A wide stretch of the mountain is circled ere they come to the gentle angle of this terrace of the proud, whose glory is tempered as a morning star, and who promises them an easier ascent henceforth (73-96). A stroke of his wing touches the poet's brow, who then approaches such a stair as was made to ease the ascent to San Miniato in the good old days when weights and measures were true and public records ungarbled (97-108). As they mount the stair the blessing of the poor in spirit falls on their ears, with sound how different from the wild laments of Hell! And Dante notes how the steep ascent seems far more easy than the level terrace of a moment back (109-120).

10

The nineteenth-century translator's paraphrase reduces the multisensuous world of Dante to a single plane of continuous narrative.

Where he says "I saw," he is actually treading on the raised effigies; for Dante, "saw" is an inclusive sensory experience.

Sassetta. THE JOURNEY OF THE MAGI. *The Metropolitan Museum of Art, New York, Bequest of Maitland F. Griggs, 1943*

10

The moment of foreshortening and realism in the image of the horse in perspective corresponds to the private narrative voice of the poet.

While the bounding outline of the profiled horses follows the object as known, the outline of the foreshortened animal follows the animal as seen.

Under the impact of perspective the seen and the known were equated, with much loss of sensory life. There is only one perspective line in the painting, at the top right. Otherwise, there is no diagonal thrust *into* the painting. Tentative indication of the new mode.

This is the kind of art that revealed the new dimensions of existence to that time. What kinds of art could serve today to probe and reveal the hidden dimensions of the electronic world?

11 KING LEAR
William Shakespeare

ACT IV, SCENE 6

GLOUCESTER: Methinks y'are better spoken.
EDGAR: Come on, sir; here's the place. Stand still. How fearful
And dizzy 'tis to cast one's eyes so low!
The crows and choughs that wing the midway air
Show scarce so gross as beetles. Halfway down
Hangs one that gathers sampire—dreadful trade!
Methinks he seems no bigger than his head.
The fishermen that walk upon the beach
Appear like mice; and yond tall anchoring bark,
Diminish'd to her cock; her cock, a buoy
Almost too small for sight. The murmuring surge
That on th' unnumb'red idle pebble chafes
Cannot be heard so high. I'll look no more,
Lest my brain turn, and the deficient sight
Topple down headlong.
GLOUCESTER: Set me where you stand.
EDGAR: Give me your hand. You are now within a foot
Of th' extreme verge. For all beneath the moon
Would I not leap upright.

Shakespeare offers a form of "lesson" in perspective by a series of five unconnected visual planes:

11

Plane 1 – "crows and choughs"
Plane 2 – "halfway down"
Plane 3 – "the fishermen"
Plane 4 – "and yond tall anchoring bark"
Plane 5 – "her cock, a buoy"

The formal perspective in *Lear* is presented as a very unpleasant experience—the breaking out of the warm, familiar multisensory spaces into fragmented visual space. This is a change from corporate to private space.

The silence of the scene intensifies the unpleasantness of the perspective, as later, with Pascal: *"Le silence éternel de ces espaces infinis m'affrait."*

Hieronymus Bosch. THE GARDEN OF DELIGHTS (DETAIL). *The Prado Museum, Madrid*

11

Bosch injected the spaces of the medieval dream world into the new Renaissance spaces. Kafka, in the twentieth century, introduced the new world of the unconscious into the old rational space of the Renaissance.

A later phase of the Bosch nightmare appears in Blake's horror at the dissolution of the unity of the imagination by way of vision and "Newton's sleep."

Vision, as our only objective and detached sense, when in high definition, discourages empathy.

THE FAERIE QUEENE
Edmund Spenser

BOOK I, CANTO I

The Patron of true Holiness,
Foule Errour doth defeate
Hypocrisie, him to entrape,
Doth to his home entreate.

A Gentle Knight was pricking on the plaine,
Ycladd in mightie armes and siluer shielde,
Wherein old dints of deepe wounds did remaine,
The cruell markes of many a bloudy fielde;
Yet armes till that time did he neuer wield:
His angry steede did chide his foming bitt,
As much disdayning to the curbe to yield:
Full iolly knight he seemd, and faire did sitt,
As one for knightly giusts and fierce encounters fitt.

The stanza presents an action in three planes, two of space and one of time, each one of which is like a single panel, or frame, of a Medieval narrative painting, where there is parataxis rather than linkage.

12

The vector of the speaker's voice tends to incline these separate planes toward each other. The preference of the age is in this direction of continuity and connectedness.

Rhyme and alliteration intensify the tactile quality of the images.

Ben Jonson said: "Spenser writ no language." Spenser's old-fashioned spelling is a façade of antiquarianism—a constraint on the new visual gradient of print and perspective.

Piero di Cosimo. VULCAN AND AEOLUS. *The National Gallery of Canada, Ottawa*

12

Di Cosimo turns the Medieval ear world into a Renaissance eye world.

The rebirth of continuous visual space after a thousand years militated against the interfaces of transparency and overlay.

"To the blind all things are sudden."—ALEX LEIGHTON

13 BARTHOLOMEW FAIR
Ben Jonson

Enter Costard-monger, followed by Nightingale.

COST. Buy any pears, pears, fine, very fine pears!
TRASH. Buy any gingerbread, gilt gingerbread!
NIGHT. Hey, now the Fair's a filling! *(Sings.)*
    O, for a tune to startle
    The birds o' the booths here billing,
    Yearly with old saint Bartle!
    The drunkards they are wading,
    The punks and chapmen trading;
    Who'd see the Fair without his lading?

    Buy any ballads, new ballads?

Cries and sounds of the fair create an orchestral togetherness.

13

The fair is an all-at-once "Happening" of multifarious events—
all the fairs that ever were.

In Jonson's conceptual, nonperspective world:
"His emotional tone is not in the single verse, but in the design
of the whole. . . . the marvel of the play is the bewildering
rapid chaotic action of the fair;
it is the fair itself, not anything that happens
to take place in the fair."—T. S. ELIOT

Eliot understood the world of difference between formal
and pictorial space.

The bounding line does not enclose anything, it *is:*
"We cannot call a man's work superficial when it is the creation
of a world; a man cannot be accused of dealing
superficially with the world which he himself has created;
the superficies *is* the world."—T. S. ELIOT

Pieter Brueghel. THE PARABLE OF THE BLIND. *National Gallery, Naples*

Counterpoint between content (the blind) and highly visual picture structure. To the sighted the blind occupy a mysterious space. What about people wearing dark glasses? **13**

Polarity between aimless facial directions and the line of motion.

Tactile linkage between figures is nonvisual. Compare "a crocodile of girls," a French phrase for "vertebrate," two-and-two perambulation.

Another tension exists between the visually organized background and the haptic foreground.

14 LE MORTE D'ARTHUR
Sir Thomas Malory

CHAPTER II

And so as Sir Mordred was at Dover with his host, there came King Arthur with a great navy of ships, and galleys, and carracks. And there was Sir Mordred ready awaiting upon his landing, to let his own father to land upon the land that he was king over. Then there was launching of great boats and small, and full of noble men of arms; and there was much slaughter of gentle knights, and many a full bold baron was laid full low, on both parties. But King Arthur was so courageous that there might no manner of knights let him to land, and his knights fiercely followed him; and so they landed maugre Sir Mordred and all his power, and put Sir Mordred aback, that he fled and all his people.

So when this battle was done, King Arthur let bury his people that were dead. And then was noble Sir Gawaine found in a great boat, lying more than half dead. When Sir Arthur wist that Sir Gawaine was laid so low, he went unto him; and there the king made sorrow out of measure, and took Sir Gawaine in his arms, and thrice he there swooned.

This eyrie vision, as it were, of a military landing lacks a private point of view. **14**

As with the Medieval map, the text is a probe rather than a chart—a marriage of eye and ear.

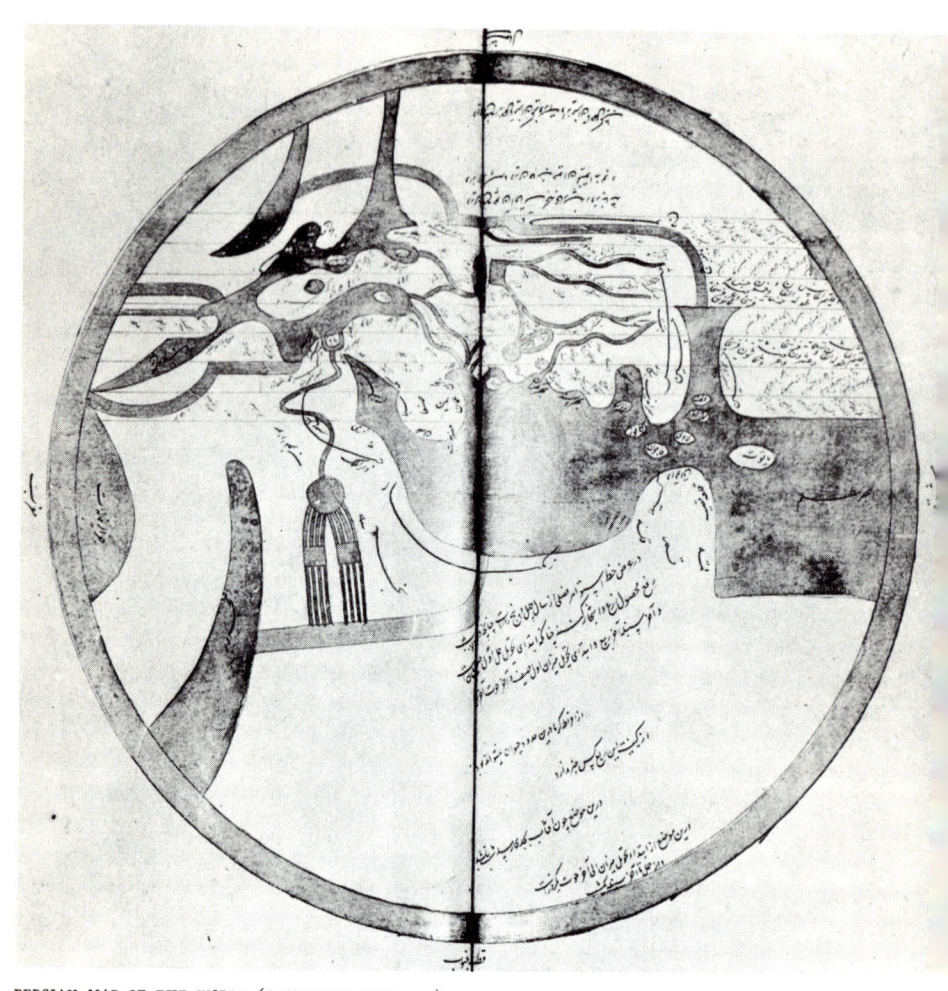

PERSIAN MAP OF THE WORLD (SIXTEENTH CENTURY)

14

Macrobius in the fourth and fifth centuries repeated the earlier geographic idea of Cicero that the world has five zones. Though likely inhabited, he says of the South Temperate Zone: "We never have had, and never shall have, the possibility of discovering by whom."

No upside down or straight lines in } Medieval maps (small child's world) / Eskimo art / Space capsule

Columbus sailed out of sacred (auditory) space into profane (visual) space by linear intent. Arrival through design.

This map is total field rather than specialist direction.

The Medieval cartographer played his space kinetically, like the Eskimo today. "The ear bone connected to the eye bone."

The eye defeated the Aborigines.

Maps could also be configurations for the cosmic powers, in the sense suggested by Gerald Hawkins:

Stonehenge makes no sense when seen from the ground. It is impressive only when seen in plan from above. But neolithic man had no airplanes from which to view his own work—therefore he may have been signalling his prowess to the powers in the sky . . . to his gods.

"So Geographers, in Afric-Maps,
With Savage-Pictures fill their Gaps;
And o'er uninhabitable Downs
Place Elephants for want of Towns. . . ."
—SWIFT

"Geography is about maps. Biography is about chaps."—G.K.C.

This map, in fact, says a lot about chaps and cultures.

15 ABSALOM AND ACHITOPHEL
John Dryden

Of these the false Achitophel was first;
A name to all succeeding ages curst:
For close designs and crooked counsels fit,
Sagacious, bold, and turbulent of wit,
Restless, unfixd in principles and place,
In powr unpleasd, impatient of disgrace;
A fiery soul which, working out its way,
Fretted the pigmy body to decay:
And o'er-informd the tenement of clay.
A daring pilot in extremity;
Pleased with the danger, when the waves went high,
He sought the storms; but, for a calm unfit,
Would steer too nigh the sands, to boast his wit.
Great wits are sure to madness near allied,
And thin partitions do their bounds divide. . . .

Couplet acts as a recessive plane immediately beside a vertical plane: *Fuga per canonem*.

15

Irrational content in a rational frame.

The integral being cannot be a hero, or even tolerated, in a fragmented or specialist society.

El Greco. THE BURIAL OF COUNT D'ORGAZ. *Church of San Tomé, Toledo*

15

Small heads cleverly yield elevation and reverence.

The figures are distorted as if on a recessive plane, then presented on frontal-parallel planes.

Object as seen yields distortion.
Object as known precludes distortion.

T. S. Eliot on Jonson's characters as conforming "to the logic of the emotions of their world. . . . They are not fancy, because they have a logic of their own; and this logic illuminates the actual world, because it gives us a new point of view from which to inspect it."
Unlike Jonson's world, El Greco's world is highly visual and fosters the "fancy."

THE COLLAR
George Herbert

I struck the board, and cried, "No more;
    I will abroad!
What, shall I ever sigh and pine?
My lines and life are free; free as the road,
  Loose as the wind, as large as store.
      Shall I be still in suit?
  Have I no harvest but a thorn
  To let me blood, and not restore
What I have lost with cordial fruit?
      Sure there was wine,
  Before my sighs did dry it; there was corn
    Before my tears did drown it;
  Is the year only lost to me?
    Have I no bays to crown it,
No flowers, no garlands gay? all blasted,
      All wasted?
  Not so, my heart; but there is fruit,
    And thou hast hands.
  Recover all thy sigh-blown age
On double pleasures; leave thy cold dispute
Of what is fit and not; forsake thy cage,
    Thy rope of sands
Which petty thoughts have made; and
      made to thee
  Good cable, to enforce and draw,
    and be thy law,
While thou didst wink and wouldst not
    see.
      Away! take heed;
      I will abroad.

Call in thy death's-head there, tie up thy
    fears;
        He that forbears
    To suit and serve his need
        Deserves his load."
But as I raved and grew more fierce and wild
    At every word,
Methought I heard one calling, "Child";
    And I replied, "My Lord!"

A peripety at the end, the moment of truth or recognition achieved by contrast. **16**

A stepping up of visual values makes a new dichotomy between the spiritual and the material.

The casual social space and colloquial histrionics dramatized against the inner voice and space.

Michelangelo Caravaggio. THE CALLING OF ST. MATTHEW. *Church of San Luigi dei Francesi, Rome*

16

The isolated moment moves us toward photographic stress on visual realism.

Darkness is to space what silence is to sound, i.e., the interval.

No ambient light—the world of the proscenium arch and stage lighting.

The contrast between low living and high thinking—the Caravaggio formula.

"To live without clocks is to live forever."—R.L.S.
Time is only divisible in visual space.

17 THE REVENGER'S TRAGEDY
Cyril Tourneur

ACT III, SCENE V

VINDICE: The very same.
And now methinks I could e'en chide myself
For doting on her beauty, though her death
Shall be revengd after no common action.
Does the silkworm expend her yellow labours
For thee? for thee does she undo herself?
Are lordships sold to maintain ladyships
For the poor benefit of a bewitching minute?
Why does yon fellow falsify highways,
And put his life between the judge's lips,
To define such a thing? keeps horse and men
To beat their valours for her?
Surely we are all mad people, and they
Whom we think are, are not; we mistake those:
'Tis we are mad in sense, they but in clothes.

Vindice uses his mistress' skull as a mirror for the entire social sphere: "To seize and clutch and penetrate, expert beyond experience."

17

A melodrama contrasting shadow and substance, inner and outer, outer appearance and inner motives?

The dilemma of visual culture is already indicated in a growing sense of division between appearance and reality: "There is no art to read the mind's construction in the face." Or, "That one may smile, and smile, and be a villain." That visual culture introduced the obsession with the problem of hypocrisy, whether in Molière's *Tartufe* or Fielding's *Tom Jones*.

Georges de La Tour. MARY MAGDALEN WITH A MIRROR. *Collection André Fabius, Paris*

17

Eyeball to eyeball. The skull is a Baroque mirror.

The visual confrontation of life as shallow horror. In a visual culture honor can assume the character of profundity.

The *feel* of the dark against the *sight* of the skull.

A universal symbol of death appears as a sensational stage property. "No contact.possible to flesh/Allayed the fever of the bone."—T. S. ELIOT, "Whispers of Immortality"

New alienation from inclusive unity by specialist fragmentation.

The mirror reflects the Baroque quest for depth through duality.

18 SONNET 73
William Shakespeare

That time of year thou mayst in me behold
When yellow leaves, or none, or few, do hang
Upon those boughs which shake against the cold,
Bare ruin'd choirs where late the sweet birds sang.
In me thou see'st the twilight of such day
As after sunset fadeth in the West,
Which by-and-by black night doth take away,
Death's second self, that seals up all in rest.
In me thou see'st the glowing of such fire
That on the ashes of his youth doth lie,
As the deathbed whereon it must expire,
Consum'd with that which it was nourish'd by.
    This thou perceiv'st, which makes thy love more strong,
    To love that well which thou must leave ere long.

The new print-created public inspired Montaigne: "I owe a complete portrait of myself to my public."

18

It had earlier inspired Machiavelli to inflate the image of the Prince.

The Shakespearean moment ("that time of year") includes several times at once just as his image of "bare ruin'd choirs" includes more than one theme and is a visual pun.

The sonneteers substituted public notice for the loss of immediate participation in the social ritual.

Rembrandt van Rijn. SELF-PORTRAIT. *The National Gallery of Canada, Ottawa*

18

A time dominated by Du Fresnoy's admonition: "Let your thoughts be wholly taken up with acquiring to yourself a glorious Name, which can never perish, but with the World; and make that the Recompense of your worthy Labours."

New self-awareness created by new public environment.

The new public as mirror reveals the private dimension, spurring the new enterprise of self-expression.

The dualism of the Baroque idea appears in the duality of the profile as it is drawn by light from one position and as it is seen from the position of the artist.

SONG
Thomas Carew

Ask me no more where Jove bestows,
When June is past, the fading rose;
For in your beauty's orient deep
These flowers, as in their causes, sleep.

Ask me no more whither do stray
The golden atoms of the day;
For in pure love heaven did prepare
Those powders to enrich your hair.

Ask me no more whither doth haste
The nightingale when May is past;
For in your sweet dividing throat
She winters and keeps warm her note.

Ask me no more where those stars 'light
That downwards fall in dead of night;
For in your eyes they sit, and there
Fixèd become as in their sphere.

Ask me no more if east or west
The Phoenix builds her spicy nest,
For unto you at last she flies,
And in your fragrant bosom dies.

Renaissance poets and painters alike made notebooks and catalogues of postures and verbal figures. **19**

Clichés and stereotypes enacted a drama of social stability.

The lady is ironically presented as a Sargasso Sea of decayed riches.

Peter Lely. PORTRAIT OF THE COUNTESS OF MEATH. *The National Gallery of Canada, Ottawa*

Lady Meath as Diana—mythic dimension created by the trick omission of middle distance both in time and space.

19

Here is the prologue to the drama of Big Brother Watching You that later unfolds in the *Tattler* and the *Spectator* (*lo spettatore nel centro del quadro*).

Aristocracy as the custodian of tradition and values.

A slice of mime in a slice of time.

Of Charles I, by Van Dyck, Henri Focillon: "The Portrait as Sitter."

THE RAPE OF THE LOCK
Alexander Pope

    Say what strange motive, Goddess! could compel
A well-bred Lord t' assault a gentle Belle?
O say what stranger cause, yet unexplored,
Could make a gentle Belle reject a Lord?
In tasks so bold, can little men engage,
And in soft bosoms dwells such mighty Rage?
    Sol through white curtains shot a timorous ray,
And oped those eyes that must eclipse the day:
Now lap dogs give themselves the rousing shake,
And sleepless lovers, just at twelve, awake:
Thrice rung the bell, the slipper knocked the ground,
And the pressed watch returned a silver sound.

Meticulous care for ritual routines permitted them to ignore the dissolution of a whole world. Note the comic inventory of the senses at the beginning of the Pope poem.

20

Pope's playfulness occurs within nursery dimensions of spontaneity.

Rear-view mirror world. For ". . . yonder all before us lie/Deserts of vast eternity."

Antoine Watteau. PERFECT HARMONY *(L'Accord Parfait). The National Trust, Waddesdon Manor, England*

"Created by a non-participant just outside the borders of life."—JOHN CANADAY

20

The great Baroque waves of Rubens are here reduced to the ripples on a satin skirt.

Chinese saying applies: "People in the West are always getting ready to live."

A playful, elegant world which can only exist in the equivalent of the circumscribed spaces of the boudoir.

21 PARADISE LOST
John Milton

BOOK II

High on a throne of royal state, which far
Outshone the wealth of Ormus and of Ind,
Or where the gorgeous East with richest hand
Showers on her kings barbaric pearl and gold,
Satan exalted sat, by merit rais'd
To that bad eminence....

Milton's fixed perspective can be extrapolated from the weight of Satan. The weight of Satan can be inferred from the fact that he is portrayed in fixed perspective.

21

Sighted Milton had little perspective in his early poems. Blind Milton increased his perspective vision steadily, climaxed in *Samson Agonistes*.

L'Allegro and *Il Penseroso* are poems by the sighted Milton, in the *Book of Hours* tradition, which catalogue the events of the day in less perspective even than Pol Limbourg **(see 9)**.

Blind Milton is perhaps the first to develop perspective by verbal means. Could his blindness have led to visual stress?

Giovanni Battista Tiepolo. CEILING FRESCO OF THE KAISERSAAL. *Wurzburg Residenz, Germany*

21

Expansive grandeur effected by pronounced recession upward.

The vertical and horizontal dualism of the steps serve to stabilize the dynamic elliptical form.

Flying figures in visual space are a contradiction in fact. Medieval angels are not sustained by air; they have no weight.

The way an object is lighted is its profile from the point of view of the sun. The Baroque artist gives up his place to the sun. You look at it. It doesn't look at you. Form: the emanating image, from cause to effect. The High Baroque is resonance at the expense of the tactile; a stress on kinesis, playing down active touch or involvement.

THE SOLITARY REAPER
William Wordsworth

Behold her, single in the field,
Yon solitary Highland Lass!
Reaping and singing by herself;
Stop here, or gently pass!
Alone she cuts and binds the grain,
And sings a melancholy strain;
O listen! for the Vale profound
Is overflowing with the sound.

No Nightingale did ever chaunt
More welcome notes to weary bands
Of travellers in some shady haunt,
Among Arabian sands:
A voice so thrilling ne'er was heard
In spring-time from the Cuckoo-bird,
Breaking the silence of the seas
Among the farthest Hebrides.

Will no one tell me what she sings?
Perhaps the plaintive numbers flow
For old, unhappy, far-off things,
And battles long ago:
Or is it some more humble lay,
Familiar matter of to-day?
Some natural sorrow, loss, or pain,
That has been, and may be again!

The alien and exotic element in the picture consists of the poet and his friends. **22**

To paraphrase Ogden Nash, carry me off to the Highlands and there you'll meet a lot of people from the Lake Country.

The beautiful is in the foreground; the sublime is given as an auditory background of resonating traditions and expeditions.

"By 1770 (the year of Wordsworth's birth) touring and sketching the lakes had become trite."
—CHRISTOPHER HUSSEY, *The Picturesque*

Thomas Gainsborough. LANDSCAPE WITH A BRIDGE. *National Gallery of Art, Washington, D.C.*

22

No leap yet from the beautiful to the sublime
—it comes a little later.

A lake "will appear to most advantage when approached from its outlet, especially if the lake be in a mountainous country."
—WORDSWORTH, *Guide to the Lakes*

Romantic poets and painters offer probes into the world of the aesthetic process.

The eighteenth century was dedicated to the proposition that the outer world existed to end in a picture.

23 LOVELIEST OF TREES
A. E. Housman

Loveliest of trees, the cherry now
Is hung with bloom along the bough,
And stands about the woodland ride
Wearing white for Eastertide.

Now, of my threescore years and ten,
Twenty will not come again,
And take from seventy springs a score,
It only leaves me fifty more.

And since to look at things in bloom
Fifty springs are little room,
About the woodlands I will go
To see the cherry hung with snow.

Housman enters the Oriental world of the Japanese print through nostalgia for the Pooh Bear world of touch.

23

Is it a dare to go down to the woods today?

Housman reverses Chiang Yee by placing visual spaces in an Oriental space, a sort of psychosemantic jest.

Chiang Yee. COWS AT DERWENTWATER. *Permission of the artist*

23

> "What does he like?
> He likes what he can paint."
> —NIETZSCHE

> "Art is the expression of an enormous preference."
> —WYNDHAM LEWIS

The meeting of West and East. Oriental icons in the unconnected vertical planes of classical visual space.

The classical mode of the fifteenth century (see Di Cosimo), like the Medieval before it, is closer to the Oriental than the Baroque is.

LONDON
William Blake

I wander through each chartered street,
Near where the chartered Thames does flow,
And mark in every face I meet
Marks of weakness, marks of woe.

In every cry of every man,
In every infant's cry of fear,
In every voice, in every ban,
The mind-forged manacles I hear;

How the chimney-sweeper's cry
Every blackening church appalls,
And the hapless soldier's sigh
Runs in blood down palace walls.

But most, through midnight streets I hear
How the youthful harlot's curse
Blasts the new-born infant's tear,
And blights with plagues the marriage hearse.

Blast and blight as natural growths of a blueprinted world.

24

The chartered, the man-made, is counterpointed with the natural world of innocence. In an earlier draft:

> "I wander thro' each dirty street,
> Near where the dirty Thames does flow,
> And mark in every face I meet,
> Marks of weakness, marks of woe."

The shift to "chartered" is a change to montage and transparency from simple description.

A shift from single-level space to multilevel space.

Pierre Jacques de Loutherbourg. IRON WORKS AT COALBROOKDALE. *From a Series of Views of British Scenery*

24

"Though every prospect pleases,
And only man is vile."
—REGINALD HEBER

Not yet the astronaut. The man-made environment usurps the natural world.

De Loutherberg substitutes the factory for the sublime. Nature imitating art?

This way to the integration of work and beauty?

25 ELEGY WRITTEN IN A COUNTRY CHURCHYARD
Thomas Gray

The curfew tolls the knell of parting day,
   The lowing herd winds slowly o'er the lea,
The plowman homeward plods his weary way,
   And leaves the world to darkness and to me.

Now fades the glimmering landscape on the sight,
   And all the air a solemn stillness holds,
Save where the beetle wheels his droning flight,
   And drowsy tinklings lull the distant folds:

Save that from yonder ivy-mantled tower
   The moping owl does to the moon complain
Of such as, wandering near her secret bower,
   Molest her ancient solitary reign.

25

The Elegy walks backward into the future.

Sentimentality, like pornography, is fragmented emotion; a natural consequence of a high visual gradient in any culture.

The funereal pomp of Gray's heroic aphorisms stems from the ancient traditions of moral wisdom. Environed by homely objects, they are unified by auditory resonance.

Benjamin West. THE DEATH OF WOLFE. *The National Gallery of Canada, Ottawa*

25

No bugle could sound across such a landscape. It exists only in a moment. Where the flat Persian image serves all times via sound (**see 6**), the three-dimensional space can exist only in a fragmented moment of time.

Acoustical space cannot exist in a fragment of visual space.

West is reportage in the style of a repertory theater.

The sublime is in the foreground, anticipating, in reverse, the later Romantic spatial organization.

26 THE RIME OF THE ANCIENT MARINER
Samuel Taylor Coleridge

"And a good south wind sprung up behind;
The Albatross did follow,
And every day, for food or play,
Came to the mariners' hollo!

"In mist or cloud, on mast or shroud,
It perched for vespers nine;
Whiles all the night, through fog-smoke white,
Glimmered the white moon-shine."

"God save thee, ancient Mariner!
From the fiends, that plague thee thus!—
Why look'st thou so?"—"With my cross-bow
I shot the Albatross!"

A marginal gloss added to the second edition of the poem yields a kind of eerie, electronic music. **26**

The ballad form was a new vortex. A corporate mask of the pastoral antique.

The Rime exploits the tension between normal festive space and resonating eerie space (the beautiful and the sublime).

Henri Füssli. THE NIGHTMARE. *Goethe-Museum, Frankfort, Germany*

A transparent overlay of the human and the superhuman.

26

The recovery of iconic and sensory involvement via horror.

> "This nonday diary,
> This allnights newseryreel"
> —JAMES JOYCE, *Finnegans Wake*

By introducing the proprioceptive-visceral into the tortured, neo-classical pose, the artist induces an empathic response.

Dream vision as escape from the dominance of rational-visual values.

Is Füssli more fanciful than imaginative?

To the spectator the horrific images are background. To the dreamer they are foreground. You can't dream pictorially but only iconically. Does the psychiatrist form story lines for dreams?

THE TYGER
William Blake

Tyger! Tyger! burning bright
In the forests of the night,
What immortal hand or eye
Could frame thy fearful symmetry?

In what distant deeps or skies
Burnt the fire of thine eyes?
On what wings dare he aspire?
What the hand dare sieze the fire?

And what shoulder, & what art,
Could twist the sinews of thy heart?
And when thy heart began to beat,
What dread hand? & what dread feet?

What the hammer? what the chain?
In what furnace was thy brain?
What the anvil? what dread grasp
Dare its deadly terrors clasp?

When the stars threw down their spears,
And water'd heaven with their tears,
Did he smile his work to see?
Did he who made the Lamb make thee?

Tyger! Tyger! burning bright
In the forests of the night,
What immortal hand or eye,
Dare frame thy fearful symmetry?

27

Resonating acoustic space.
A vast echo chamber for reader participation.

This tiger is not *in* any tank or any zoo. It is a world.
The symbolic does not refer—it *is*.

His symbolic tiger is structured by a catalogue of unanswered questions. A necessary artistic means to the iconic end? Involvement by suggestion rather than statement?

Blake was not attracted into the world of Romantic painting and poetry where landscape was used as a direct instrument of separate and isolated nuances of feeling and emotion. His plastic and pictorial work demanded a concert of the senses, and so did his poetry with its compression and aphoristic wit. He was really concerned with the reader, and he became completely involved in the texture of language, not in just its semantic meanings. His efforts in this regard were mostly unnoticed until the later nineteenth century. The symbolists, also, began to use language, not as a package of prepared messages, but as a heuristic probe into new experience. Prepared by their work, W. B. Yeats encountered Blake as a revolutionary experience. Blake will serve to remind us that a considerable interval of poetry and painting is to intervene between his discoveries and their further development by people like Rimbaud, Mallarmé, Eliot and Joyce.

The Tyger.

Tyger Tyger, burning bright,
In the forests of the night;
What immortal hand or eye,
Could frame thy fearful symmetry?

In what distant deeps or skies.
Burnt the fire of thine eyes?
On what wings dare he aspire?
What the hand, dare sieze the fire?

And what shoulder, & what art,
Could twist the sinews of thy heart?
And when thy heart began to beat,
What dread hand? & what dread feet?

What the hammer? what the chain,
In what furnace was thy brain?
What the anvil? what dread grasp,
Dare its deadly terrors clasp!

When the stars threw down their spears
And water'd heaven with their tears;
Did he smile his work to see?
Did he who made the Lamb make thee?

Tyger Tyger burning bright,
In the forests of the night;
What immortal hand or eye,
Dare frame thy fearful symmetry?

William Blake. "THE TYGER" (PAGE FROM THE ORIGINAL EDITION OF *Songs of Experience*).
The Pierpont Morgan Library, New York

The bounding line above all as a counterstimulant for the senses in an age of jaded pictorialism.

27

Blake, in this day, would have preferred the comic book to the photograph?

Perhaps you would prefer a Delacroix tiger to this icon?

Visual sensory fragmentation scorned. Iconic sensory unity used as exploratory thrust into new age.

Blake's insistence on art as a means of perception appears in his own aphorism:

"No man can embrace True Art
Until he has Explor'd and cast out False Art."

Blake, the craftsman, fought industrial specialism and fragmentation, by writing, designing and engraving his own works.

SHE WALKS IN BEAUTY
Lord Byron

She walks in beauty, like the night
   Of cloudless climes and starry skies;
And all that's best of dark and bright
   Meet in her aspect and her eyes:
Thus mellowed to that tender light
   Which heaven to gaudy day denies.

One shade the more, one ray the less,
   Had half impaired the nameless grace,
Which waves in every raven tress,
   Or softly lightens o'er her face;
Where thoughts serenely sweet express
   How pure, how dear, their dwelling-place.

And on that cheek, and o'er that brow,
   So soft, so calm, yet eloquent,
The smiles that win, the tints that glow,
   But tell of days in goodness spent,
A mind at peace with all below,
   A heart whose love is innocent!

28

Between beauty and the stars, no middle distance, the pure Romantic formula—beauty and the sublime immediately juxtaposed.

Like Blake's tiger, she is not contained. She *is* beauty.

Compare and contrast Wordsworth's "Lucy": "A violet by a mossy stone."

The poem is in the genre of "instructions to a painter"?

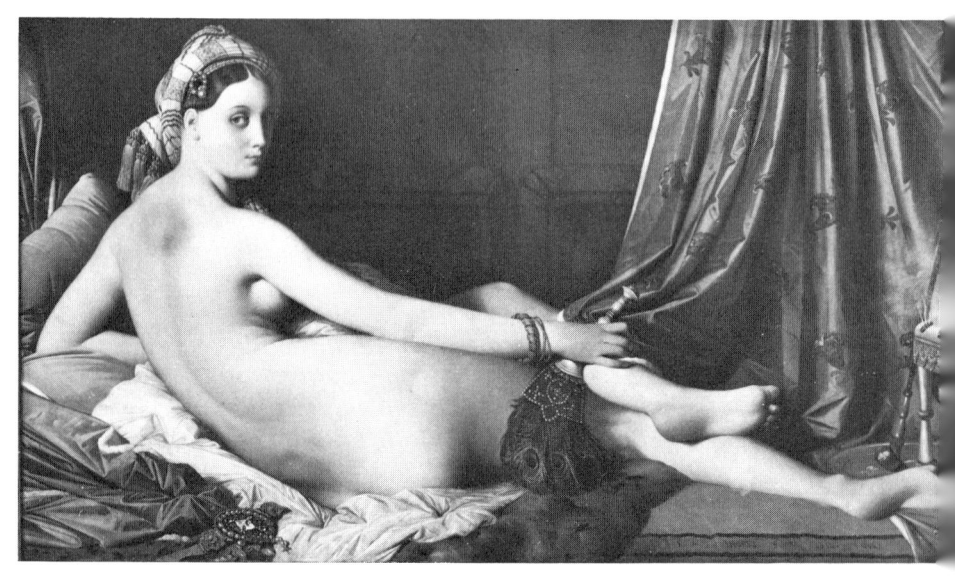

Jean-Auguste Dominique Ingres. ODALISQUE. *The Louvre, Paris*

"This beautie without falshood fayre,
Needs nought to cloath it but the ayre."
—BEN JONSON

28

Active touch evoked by kinetic sweep of unbroken line.

English law confers aesthetic status on the immobile nude,
seeing it as timeless.

A striking wariness of the head countering
the repose of the body.

The sculptural qualities of the image dim down
the purely personal identity.

THE LADY OF SHALOTT
Alfred Tennyson

Under tower and balcony,
By garden-wall and gallery,
A gleaming shape she floated by,
Dead-pale between the houses high,
    Silent into Camelot.
Out upon the wharfs they came,
Knight and burgher, lord and dame,
And round the prow they read her name,
    The Lady of Shalott.

Who is this? and what is here?
And in the lighted palace near
Died the sound of royal cheer;
And they cross'd themselves for fear,
    All the knights at Camelot:
But Lancelot mused a little space;
He said, "She has a lovely face;
God in His mercy lend her grace,
    The Lady of Shalott."

Try out the Tennyson poem as instructions for making a Rossetti painting. **29**

Only lateral space created by boat traversing the stage against a painted curtain.

Ornaments for a mood or structural vision?

Dante Gabriel Rossetti. SIR GALAHAD. *The Tate Gallery, London*

29

The use of perspective from the canvas front to the first plane is an involuntary obeisance to the continuing reign of Renaissance perception of space.

Many poets and painters strove to achieve Medieval simulation but were thwarted by fragmented sensibility.

As in a museum, direct conflict between imported artifacts and story-line showcase.

30 | THE JOLLY BEGGARS
Robert Burns

CHORUS

A fig for those by law protected!
    Liberty's a glorious feast!
Courts for cowards were erected,
    Churches built to please the priest.

AIR

What is title? what is treasure?
    What is reputation's care?
If we lead a life of pleasure,
    'Tis no matter how or where!
        A fig, etc.

With the ready trick and fable
    Round we wander all the day;
And at night, in barn or stable,
    Hug our doxies on the hay.
        A fig, etc.

The catalogue steps out of the frame of pictorial space.

30

The paradoxical sublimity of the outcast or dropout.

The plowboy poet pushed English poetry toward the oral tradition as Jazz did in the twenties.

Jolly Beggars as Romantic Beatniks. Compare the boy in the American ballad:

> "Who wandered in comfort and joy
> Into woods where the waterfalls run"

as an insight into the new Romantic curriculum—Nature as teaching machine.

> "One impulse from a vernal wood
> May teach you more of man,
> Of moral evil and of good,
> Than all the sages can."

Housman is even more iconic:

> ". . . Malt does more than Milton can
> To justify God's ways to man."

George Morland. A HILL-SIDE WITH TRAMPS REPOSING. *The National Gallery of Canada, Ottawa*

"A fig for those by law protected."

30

Romantic unenclosed space as rebellion against legally constituted spaces. Counterattack on the "shades of the prisonhouse."

A stage set for the *Beggar's Opera* (Newgate Pastoral) and its hero Captain Macheath.

"High mountains are a feeling, but the hum
Of human cities, torture."
—BYRON

The distant view, social or geographic, as a short cut to involvement and unified dream-life.

The paradox of Morland: the attempt to depict unenclosed spaces by techniques suited to pictorial enclosure.

HYMN TO INTELLECTUAL BEAUTY
Percy Bysshe Shelley

The awful shadow of some unseen Power
    Floats tho' unseen among us; visiting
    This various world with as inconstant wing
As summer winds that creep from flower to
    flower.
Like moonbeams that behind some piny mountain
    shower,
    It visits with inconstant glance
    Each human heart and countenance;
Like hues and harmonies of evening,
    Like clouds in starlight widely spread,
    Like memory of music fled,
    Like aught that for its grace may be
Dear, and yet dearer for its mystery.

This "Newton among poets" probes
psychic and aesthetic processes.

31

Nature becomes a mystery and a challenge
as much as a picture.

Could Füssli have carried out Shelley's instructions for a
painting of the "unseen Power" better than Turner?

Is Turner better than Shelley at suggesting
the "unseen Power"?

How much painterly knowledge is assumed
by the poet from the reader?

Joseph Mallord William Turner. STEAMER IN A SNOWSTORM. *The National Gallery, London*

An augury of the Romantic shift from the picturesque scene to the dynamics of process. **31**

"Painting is a science . . . an inquiry into the laws of nature."
—CONSTABLE

The key to the theme of Ruskin's "innocent eye"—pure discovery. Child's awareness as "scientific" probe, a fidelity to the natural process of knowing and learning.

A slight foretaste of light through.

Preview of the spaces of cinema, just as Seurat anticipates the spaces of TV.

The struggle of American education to obtain the freedom and spontaneity of nonvisual values by visual blueprints. Compare the dilemma of the tourist attempting to enjoy the wilderness via rail, road and photography, the exclusion of all but the visual faculty.

L'INVITATION AU VOYAGE
Charles Baudelaire

There is a wonderful country, a country of Cocaigne, they say, that I dream of visiting with an old love. A strange country lost in the mists of the North and that might be called the East of the West, the China of Europe, so freely has a warm and capricious fancy been allowed to run riot there, illustrating it patiently and persistently with an artful and delicate vegetation.

A real country of Cocaigne where everything is beautiful, rich, honest, and calm; where order is luxury's mirror; where life is unctuous and sweet to breathe; where disorder, tumult, and the unexpected are shut out; where happiness is wedded to silence; where even the cooking is poetic, rich, and yet stimulating as well; where everything, dear love, resembles you.

You know that feverish sickness which comes over us in our cold despairs, that nostalgia for countries we have never known, that anguish of curiosity? There is a country that resembles you, where everything is beautiful, rich, honest and calm, where fancy has built and decorated an Occidental China, where life is sweet to breathe, where happiness is wedded to silence. It is there we must live, it is there we must die.

Yes, it is there we must go to breathe, to dream, and to prolong the hours in an infinity of sensations. A musician has written *L'Invitation à la valse*; who will write *L'Invitation au voyage* that may be offered to the beloved, to the chosen sister?

Yes, in such an atmosphere it would be good to live— where there are more thoughts in slower hours, where clocks strike happiness with a deeper, a more significant solemnity.

Contrasted with Watteau's daintiness, Baudelaire's synesthesia is a Wagnerian uproar. **32**

The sensory life as well as the geographic is presented by polarities. Hence the haunting *déjà vu* effect.

Like Blake, Baudelaire, as the enemy of his age, saw himself as the creator of counterenvironments, as a means of perceiving the actual and invisible environment.

Eugène Delacroix. ARAB RIDER ATTACKED BY LION. *The Art Institute of Chicago*

32 Lewis Carroll's intersensory drama, "reeling and writhing and rhythmatic"?

Ouroborus—the snake swallowing its tail.
Action enclosed within the circle as a unit.

Light itself interprets the action, anticipating the Impressionists.

A sensory plunge into Africa out of the visual European frame, just as Picasso initially used the primitive as subject matter gradually assimilating its multisensory modes to his stylistic bent.

33

quite dull and stupid for things to go on in the common way

So she set to work, and very soon finished off the cake

* * * * *

"Curiouser and curiouser!" cried Alice, (she was so surprised, that she quite forgot how to speak good English,) "now I'm opening out like the largest telescope that ever was! Goodbye, feet!" (for when she looked down at her feet, they seemed almost out of sight, they were getting so far off,) "oh, my poor little feet, I wonder who will put on your shoes and stockings for you now, dears? I'm sure I can't! I shall be a great deal too far off to bother myself about you: you must manage the best way you can— but I must be kind to them," thought Alice, "or perhaps they won't walk the way I want to go! Let me see: I'll give them a new pair of boots every Christmas."

And she went on planning to herself how she would manage it.

11

Lewis Carroll. FACSIMILE PAGE FROM THE ORIGINAL MANUSCRIPT OF *Alice's Adventures Under Ground. Rare Book Division, New York Public Library*

As with Blake's handwritten texts, Carroll's makes an easy liaison with the spoken word. **33**

The shift from handwriting to print motivates some tidying up of oral diction. Contrast published edition with facsimile.

No need of story line as Carroll leaps from metamorphosis to metamorphosis.

CHAPTER II.

THE POOL OF TEARS.

"Curiouser and curiouser!" cried Alice (she was so much surprised, that for the moment she quite forgot how to speak good English); "now I'm opening out like the largest telescope that ever was! Good-bye, feet!" (for when she looked down at her feet, they seemed to be almost out of sight, they were getting so far off) "Oh, my poor little feet, I wonder

John Tenniel. PAGE FROM *Alice's Adventures in Wonderland* BY LEWIS CARROLL. *Rare Book Division, New York Public Library*

Making, not matching, is the cartoon and icon process.

33

Matching presumes to refer to outer fact; making captures inner fact.

Carroll's image is elongated throughout. She lives in a world that she makes. Tenniel, by the use of cast shadows, frames iconic Alice in three-dimensional visual space.

Carroll's drawing is integrated with the text. Tenniel's drawing (for *Punch*) is additive.

DOVER BEACH
Matthew Arnold

The sea is calm to-night.
The tide is full, the moon lies fair
Upon the straits;—on the French coast the light
Gleams and is gone; the cliffs of England stand,
Glimmering and vast, out in the tranquil bay.
Come to the window, sweet is the night-air!
Only, from the long line of spray
Where the sea meets the moon-blanched land,
Listen! you hear the grating roar
Of pebbles which the waves draw back, and fling,
At their return, up the high strand,
Begin, and cease, and then again begin,
With tremulous cadence slow, and bring
The eternal note of sadness in.

A meeting of sight and sound. **34**

Discrete "shots" of the opening lines are designed with the same tactile care for the interval as appears in the painting.

Sadness engendered by the slow rhythm of the breakers. Is it helpful for the reader to think of the oncoming waves as frontal-parallel? Or diagonal?

James Abbott McNeill Whistler. ARRANGEMENT IN BLACK AND GRAY (THE ARTIST'S MOTHER). The Louvre, Paris

Meticulous spacing on frontal-parallel planes proclaims the recent encounter with Japanese prints. **34**

The art of the profile automatically rejects the public. *"Hypocrite lecteur, mon semblable, mon frère."* Baudelaire recognizes that the reader is totally participant in the poetic process—is not a passive spectator.

Whistler's title *Arrangement in Black and Gray* has ironically been personalized into *Whistler's Mother*—the public was not ready for "arrangements."

35 AFTER THE FLOOD
Arthur Rimbaud

As soon as the idea of the Flood had subsided,

A hare stopped in the clover and the swinging flower-bells, and said its prayer through the spider's web to the rainbow.

The precious stones were hiding, and already the flowers were beginning to look up.

The butchers' blocks rose up in the dirty main street, and ships were pulled out toward the sea, piled high as in pictures.

Blood flowed in Blue Beard's hours, in the slaughter houses, in the circuses, where the seal of God whitened the windows. Blood and milk flowed.

Beavers set about building. Coffee urns let out smoke in the bars.

In the large house with windows still wet, children in mourning looked at exciting pictures.

A door slammed. On the village square the child swung his arms around; and was understood by the weather vanes and the steeple cocks everywhere, under the pelting rain.

Madame X installed a piano in the Alps. Mass and first Communions were celebrated at the 100,000 altars of the Cathedral.

The caravans departed. And the Hôtel Splendide was built in the chaos of ice and polar night.

Playing with the sensory life as one might play with the stars in a planetarium. **35**

The new world as palimpsest.

A world of disconnected spaces and times.

Movie set for Ravel's *L'Enfant et les Sortilèges*.

Henri Rousseau. THE DREAM. *Collection, The Museum of Modern Art, New York, Gift of Nelson A. Rockefeller*

Nature appears as a mechanized artifact. A machine-tooled nature for the industrial age. **35**

> *"Where the Gauguin maids*
> *In the banyan shades*
> *Wear palmleaf drapery"*
> —T. S. ELIOT, "Fragment of an Agon"

The natural habitat here presented as an artifact points to the electronic age.

$$\text{Afterimage for the world of Delacroix} = \begin{cases} \text{anti-visual} \\ \text{anti-kinetic} \\ \text{anti-sound} \end{cases}$$

HYSTERIA

T. S. Eliot

As she laughed I was aware of becoming involved in her laughter and being part of it, until her teeth were only accidental stars with a talent for squad-drill. I was drawn in by short gasps, inhaled at each momentary recovery, lost finally in the dark caverns of her throat, bruised by the ripple of unseen muscles. An elderly waiter with trembling hands was hurriedly spreading a pink and white checked cloth over the rusty green iron table, saying: 'If the lady and gentleman wish to take their tea in the garden, if the lady and gentleman wish to take their tea in the garden . . .' I decided that if the shaking of her breasts could be stopped, some of the fragments of the afternoon might be collected, and I concentrated my attention with careful subtlety to this end.

Rationality gone mad. **36**

Like the road in Munch, the opening simulation of analytic prose polarizes the world.

The waiter is an injection of the rational into an irrational situation. A counterenvironment which catapults the hysterical situation to a significant level.

Edvard Munch. THE SCREAM. *National Gallery, Oslo*

Undulating swirl of landscape pulsates with the scream.
A proprioceptive universe. **36**

The "rational" road, as an orientation point,
amplifying the scream.

DUNS SCOTUS'S OXFORD
Gerard Manley Hopkins

Towery city and branchy between towers;
Cuckoo-echoing, bell-swarmèd, lark-charmèd rook-racked,
  river-rounded;
The dapple-eared lily below thee; that country and town did
Once encounter in, here coped and poisèd powers;

Thou hast a base and brickish skirt there, sours
That neighbour-nature thy grey beauty is grounded
Best in; graceless growth, thou hast confounded
Rural rural keeping—folk, flocks, and flowers.

Yet ah! this air I gather and I release
He lived on; these weeds and waters, these walls are what
He haunted who of all men most sways my spirits to peace;

Of realty the rarest-veinèd unraveller; a not
Rivalled insight, be rival Italy or Greece;
Who fired France for Mary without spot.

37

Pattern achieved by minimizing syntax intervals (touch).

"Sprung rhythm."

"Design, pattern or what I call inscape is what I aim at in poetry."

Nature as light source (inscape) provides reader with direct spiritual message.

Hopkins anticipates the electronic age in perceiving nature itself as an art form.

Georges Seurat. A SUNDAY AFTERNOON ON THE ISLAND OF LA GRANDE JATTE. *The Art Institute of Chicago*

Painting as light source flips viewer into vanishing point. **37**

The Oriental moment of reversal—
Seurat prophet of TV.

Seurat is the art fulcrum between Renaissance visual and modern tactile. The coalescing of inner and outer, subject and object.

Seurat, by divisionism, anticipates quadricolor reproduction and color TV.

Foreshortening as one adjunct of perspective is not relevant in "light through" situation.

A GAME OF CHESS

T. S. Eliot

When Lil's husband got demobbed, I said—
I didn't mince my words, I said to her myself,
Hurry up please its time
Now Albert's coming back, make yourself a bit smart.
He'll want to know what you done with that money he gave
  you
To get yourself some teeth. He did, I was there.
You have them all out, Lil, and get a nice set,
He said, I swear, I can't bear to look at you.

On the social level the masterful use of colloquial cliché as corporate mask to archetypalize the members of the private party.

38

The verbal equivalent of the mask is rhythm and measure. A new poetic, or musical, rhythm is a vortex of corporate energy.

Ballet is not only a social mask, but as a method of organizing human energies it must avoid pictorial space in favor of frontal plane in order to maximize audience participation. See Eliot on poetic rhythm as social mask.

James Ensor. INTRIGUE. *Royal Museum of Fine Arts, Antwerp*

On one level the masquerade party.
On another level, the reflection of human propensities
and mental postures.

38

Mask wearer as role player wears the audience, manifesting human victimization by social forces.

Frontal-parallel planes.

The mask, like the sideshow freak, is not so much pictorial as participatory in its sensory appeal.

CHANSON INNOCENT
e. e. cummings

               I

in Just-
spring   when the world is mud-
luscious the little
lame balloonman

whistles   far   and wee

and eddieandbill come
running from marbles and
piracies and it's
spring

when the world is puddle-wonderful

the queer
old balloonman whistles
far   and   wee
and bettyandisbel come dancing

from hop-scotch and jump-rope and

it's
spring
and
    the

       goat-footed

balloonMan   whistles
far
and
wee

39

A myth is as good as a smile. An alphabetic ballet of words in rite order—a dramatic order of language as jester.

A jitterbug interface before the monolithic space of the Frug and Watusi with their sculptural inclusiveness.

Emphatic rejection of the bureaucratic modes of the Establishment.

The balloonman satyrized for the sake of the Rites of Spring?

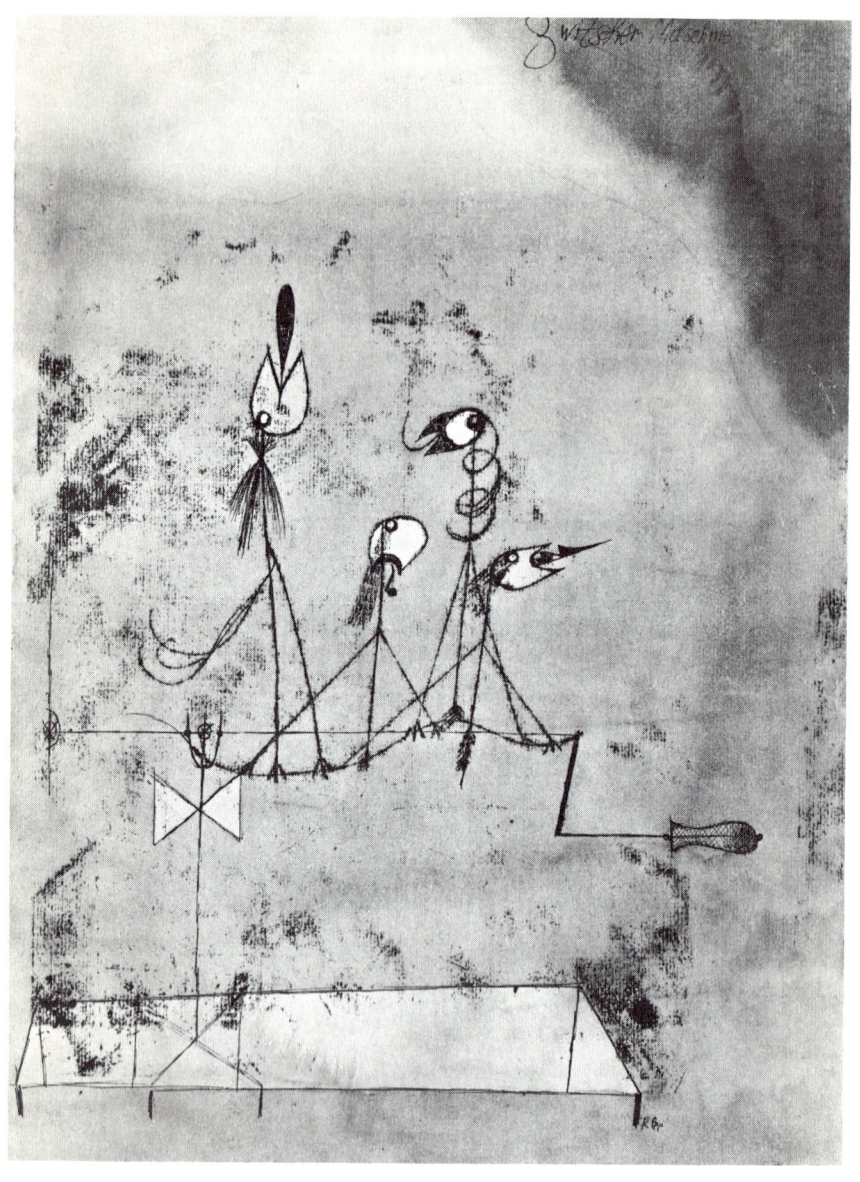

Paul Klee. THE TWITTERING MACHINE. *Collection, The Museum of Modern Art, New York*

Preview of the TV aerial, electric configuration patterned to pick up nonvisual energy. Compare "Weed in the water am I." **39**

"Abstract" art signaled the end of visual space.

The *Twittering Machine is* a kind of *sensus communis.*

PORTRAIT D'UNE FEMME
Ezra Pound

Your mind and you are our Sargasso Sea,
London has swept about you this score years
And bright ships left you this or that in fee:
Ideas, old gossip, oddments of all things,
Strange spars of knowledge and dimmed wares of price.
Great minds have sought you—lacking some one else.
You have been second always. Tragical?
No. You preferred it to the usual thing:
One dull man, dulling and uxorious,
One average mind—with one thought less, each year.
Oh, you are patient. I have seen you sit
Hours, where something might have floated up.
And now you pay one. Yes, you richly pay.
You are a person of some interest, one comes to you
And takes strange gain away:
Trophies fished up; some curious suggestion;
Fact that leads nowhere; and a tale or two,
Pregnant with mandrakes, or with something else
That might prove useful and yet never proves,
That never fits a corner or shows use,
Or finds its hour upon the loom of days. . . .

40 Pound's "Aphrodite of the Water Hole" is a perfect companion piece for Lewis' *Dreadnought of the Waterways.*

Wyndham Lewis regarded Pound and Joyce as a Sargasso Sea, a vortex of historical debris.

The individual as a montage of loosely assembled parts.

Puppet, not as extension of man, but as immediate manifestation of energy.

Wyndham Lewis. PORTRAIT OF AN ENGLISHWOMAN. *Permission of Ann Wyndham Lewis*

It might well be Miss Prism (a useful index to the fragmented space of the highly literate governing class of the Victorians) right out of Oscar Wilde.

40

The sitter's mask as a vortex is a processing of personal energy by the new industrial environment.

Sitter as puppet or servomechanism of the environment.

An assembly of environmental materials: books as head—gun turret as eyes, nose and throat.

The ape head as simulator of art activity. Compare *Apes of God:* "Playing the sedulous ape."

The insect head (world of the bridge and the machine).

Miss Britannia, 1914, models the industrial costume, an art form.

POEM IN OCTOBER
Dylan Thomas

It was my thirtieth year to heaven
Woke to my hearing from harbour and neighbour wood
    And the mussel pooled and the heron
        Priested shore
      The morning beckon
With water praying and call of seagull and rook
And the knock of sailing boats on the net webbed wall
    Myself to set foot
        That second
In the still sleeping town and set forth.

My birthday began with the water-
Birds and the birds of the winged trees flying my name
    Above the farms and the white horses
        And I rose
      In rainy autumn
And walked abroad in a shower of all my days.

He skin-dives into the ocean of his mind to
create a multisensuous image of all the birthdays
that ever were.

41

Thomas' poetry is a chamber simultaneously
echoing with many times.

Quite appropriately for the electronic age,
Dylan creates and inhabits a world sensorium.

Marc Chagall. I AND THE VILLAGE. Collection, The Museum of Modern Art, New York, Mrs. Simon Guggenheim Fund

41

Rimbaud's *avant-garde* world of montage and transparency had become the old environment by 1911.

The "innocent eye" of Chagall (nursery à gogo) stems directly from the nonpictorial spaces of the Russian village.

"We are all the primitives of a new civilization."—BOCCIONI

Any indications of revolutionary advance since then?

In the all-at-onceness of electric technology we humbly encounter the man from the backward country as *avant-garde*.

42 FIRST FAMILIES, MOVE OVER!
Ogden Nash

Carry me back to Ole Virginny,
And there I'll meet a lot of people from New York,
There the Ole Marsa of the Hounds is from Smithtown
    or Peapack or Millbrook,
And the mocking bird makes music in the sunshine
    accompanied by the rattling shaker and the
    popping cork.

The feminine world of family trees
now propped up by Madison Avenue billboards.   **42**

The spoken language itself gives immediate access
to the clichés of accepted suburbanity.

An auditory space of colloquial
idiom overlaid by a visual mosaic of stock quotations
and baseball scores.

Saul Steinberg. DRAWING. *Permission of the artist*

An elegant and "crafty" invasion of the banal as a special form of irony. **42**

Three varieties of space appear simultaneously: Renaissance, commercial-graphic and the naïf.

With entrepreneurial virtuosity Steinberg orchestrates spaces.

LEDA AND THE SWAN
William Butler Yeats

A sudden blow: the great wings beating still
Above the staggering girl, her thighs caressed
By the dark webs, her nape caught in his bill,
He holds her helpless breast upon his breast.

How can those terrified vague fingers push
The feathered glory from her loosening thighs?
And how can body, laid in that white rush,
But feel the strange heart beating where it lies?

A shudder in the loins engenders there
The broken wall, the burning roof and tower
And Agamemnon dead.
                      Being so caught up,
So mastered by the brute blood of the air,
Did she put on his knowledge with his power
Before the indifferent beak could let her drop?

A mythic, or all-at-once, vision of the beginning, middle and end of an historic cycle (the Trojan War).

43

Like Spencer, Yeats includes the audience in the ritual by making it happen now.

Yeats' realism is symbolic—Leda and her swan are not the components of some other world, but are themselves a world.

Stanley Spencer. SWAN UPPING. *The Tate Gallery, London*

43

What the Pre-Raphaelites did unconvincingly in illustrations of Medieval legend, Spencer does naturally by transcendentalizing the everyday.

Evoking ritual from the routine is to summon the muses (the daughters of memory).

Spencer summons the archetypal from the cliché.

44

The grainy sand had gone from under his feet. His boots trod again a damp crackling mast, razorshells, squeaking pebbles, that on the unnumbered pebbles beat, wood sieved by the shipworm, lost Armada. Unwholesome sandflats waited to suck his treading soles, breathing upward sewage breath. He coasted them, walking warily.—JAMES JOYCE, *Ulysses*

He reached a stretch of foreshore, uncovered by the tide, running out to a jutting headland and a misty horizon of sea. Levels of sand lay between ledges and ridges of rock. The sand sloped to the rim of the tide; there were broad plateaux of rock slippery with seaweed and bladderwrack, and opposite a gap in the cliff a rough roadway for carts and boats.
—DENYS THOMPSON, *Reading and Discrimination*

The first writer involves all the senses.

44

The second writer specializes in a visual point of view.

Which is closer to the Braque painting?

Trying to define the visual imagination, Eliot cites Macbeth:

> "Light thickens, and the crow
> Makes wing to the rocky wood,"

observing that it offers not only "something to the eye, but, so to speak, to the common sense."

The *sensus communis* as the interplay of all the senses creates an involvement that unifies the imaginative life in the way sought by William Blake.

Georges Braque. STILL LIFE: THE TABLE. *National Gallery of Art, Washington, D.C., Chester Dale Collection*

Rediscovery of primitive values in everyday objects. **44**

A point of view is a serious liability in approaching this canvas.

A recovery of the world of the cave painters.

Would this make a good TV image?

Compare it with the cave painting (**see 1**), also with Blake (**see 27**).

45 THE GREAT LOVER
Rupert Brooke

These I have loved:
    White plates and cups, clean-gleaming,
Ringed with blue lines; and feathery, faery dust;
Wet roofs, beneath the lamp-light; the strong crust
Of friendly bread; and many-tasting food;
Rainbows; and the blue bitter smoke of wood;
And radiant raindrops couching in cool flowers;
And flowers themselves, that sway through sunny hours,
Dreaming of moths that drink them under the moon;
Then, the cool kindliness of sheets, that soon
Smooth away trouble; and the rough male kiss
Of blankets; grainy wood; live hair that is
Shining and free; blue-massing clouds; the keen
Unpassioned beauty of a great machine;
The benison of hot water; furs to touch;
The good smell of old clothes; and other such—
The comfortable smell of friendly fingers,
Hair's fragrance, and the musty reek that lingers
About dead leaves and last year's ferns. . . .

> Brooke's good taste is able to avoid any fresh awareness. His inventory is charged with "friendly" and reassuring features.

45

> A generation earlier G. M. Hopkins had made an inventory of familiar items seen in unfamiliar and discontinuous aspects:

PIED BEAUTY

Glory be to God for dappled things—
    For skies as couple-colour as a brinded cow;
        For rose-moles all in stipple upon trout that swim;
Fresh-firecoal chestnut-falls; finches' wings;
    Landscapes plotted and pieced—fold, fallow, and plough;
        And áll trádes, their gear and tackle and trim.

All things counter, original, spare, strange;
    Whatever is fickle, freckled (who knows how?)
        With swift, slow; sweet, sour; adazzle, dim;
He fathers-forth whose beauty is past change:
        Praise him.

> Hopkins offers new perception related to the new world that flowed across the old one.

Be choosy.

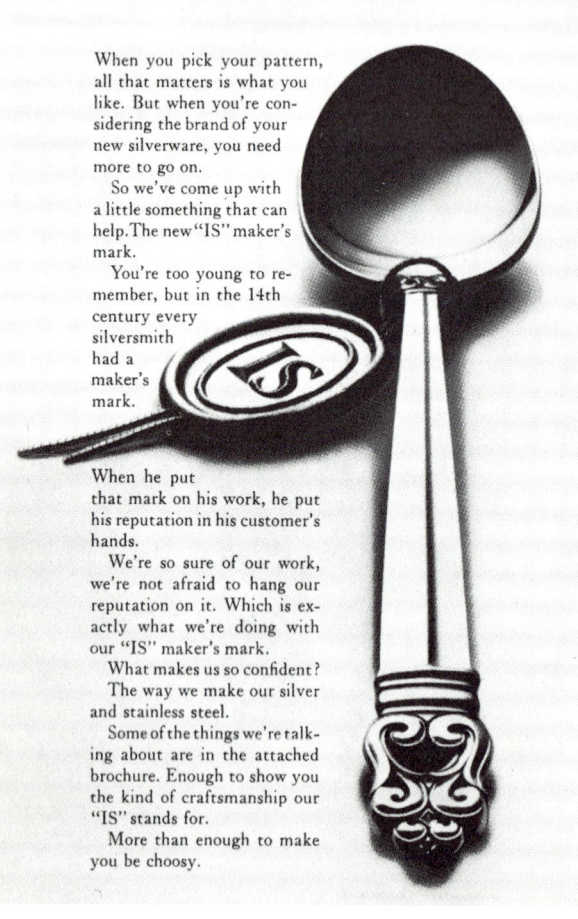

When you pick your pattern, all that matters is what you like. But when you're considering the brand of your new silverware, you need more to go on.

So we've come up with a little something that can help. The new "IS" maker's mark.

You're too young to remember, but in the 14th century every silversmith had a maker's mark.

When he put that mark on his work, he put his reputation in his customer's hands.

We're so sure of our work, we're not afraid to hang our reputation on it. Which is exactly what we're doing with our "IS" maker's mark.

What makes us so confident? The way we make our silver and stainless steel.

Some of the things we're talking about are in the attached brochure. Enough to show you the kind of craftsmanship our "IS" stands for.

More than enough to make you be choosy.

1967 ADVERTISEMENT. *The International Silver Company, Meriden, Connecticut*

45

Good taste is a sin of omission. It leaves out direct awareness of forms and situations.

Good taste is the first refuge of the noncreative. It is the last-ditch stand of the artist.

Good taste is the anesthetic of the public. It is the critic's excuse for lack of perception.

Good taste is the expression of a colossal incompetence. It is the "putting on" of the genteel audience as a mask or net by which to capture ambient snob appeal.

Good taste is the most obvious resource of the insecure. People of good taste eagerly buy the Emperor's old clothes.

Good taste is the highly effective strategy of the pretentious.

46 AN IRISH AIRMAN FORESEES HIS DEATH
William Butler Yeats

I know that I shall meet my fate
Somewhere among the clouds above;
Those that I fight I do not hate,
Those that I guard I do not love;
My country is Kiltartan Cross,
My countrymen Kiltartan's poor,
No likely end could bring them loss
Or leave them happier than before.
Nor law, nor duty bade me fight,
Nor public men, nor cheering crowds,
A lonely impulse of delight
Drove to this tumult in the clouds;
I balanced all, brought all to mind,
The years to come seemed waste of breath,
A waste of breath the years behind
In balance with this life, this death.

IN MEMORY OF W. B. YEATS
W. H. Auden

He disappeared in the dead of winter:
The brooks were frozen, the airports almost deserted,
And snow disfigured the public statues;
The mercury sank in the mouth of the dying day.
O all the instruments agree
The day of his death was a dark cold day.

Far from his illness
The wolves ran on through the evergreen forests,
The peasant river was untempted by the fashionable quays....

> "Auden's bare, upland regions provide a particular view of the world below, a panoramic view, a view of the hawk, or the airman."
> —RICHARD HOGGART, *Auden: An Introductory Essay*

46

Harley Parker. THE TRIP. *Collection of the artist*

46

A Chagall-like memory of children's games. Air flight involves an extension of the whole body. Once in the air a plane makes its own times and spaces, or perhaps one should say that it exists mainly in the dimension of time rather than space once it is off the ground. The passengers develop a "destination syndrome," as it were, as their contribution to the unique space created in the act of flight.

The picture presents a mother and her eight children at twilight. The frontal figure is still earthbound. As in the Yeats poem, there is no point of view, but rather total involvement. The painting evokes nostalgia for the imaginative and unfettered world of childhood.

Simulating the action of the technologies in our environment is one of the major thrills of childhood in North America. The machine has provided the poetry of our childhood for more than a century. Perhaps the airplane, which seems to mark the end of the regime of the wheel, exists somewhere between the machine and the human body.

A CONEY ISLAND OF THE MIND
Lawrence Ferlinghetti

    Kafka's Castle stands above the world
   like a last bastille
      of the Mystery of Existence
Its blind approaches baffle us
       Steep paths
  plunge nowhere from it
      Roads radiate into air
like the labyrinth wires
      of a telephone central
thru which all calls are
       infinitely untraceable
   Up there
     it is heavenly weather
Souls dance undressed
      together
  and like loiterers
      on the fringes of a fair
we ogle the unobtainable
      imagined mystery
    Yet away around on the far side
      like the stage door of a circus tent
is a wide wide vent in the battlements
       where even elephants
         waltz thru

One of the features of this poem is its environmental-like quality. When a poet puts aside the narrative pattern of discourse, it is natural to dwell on environments. Witness *The Waste Land* or *Finnegans Wake*. An environment is unclassifiable in a sense. Perhaps it is too multisensuous to afford any simple pictorial experience. This poem has much in common with a newspaper page, offering numerous perspectives rather than a point of view. However, Kafka's Castle serves as a sort of date line for this newspaper coverage of the world. It also serves as an image of the labyrinth with its accompanying association of the Minotaur, symbol of the encounter with the self. Kafka evokes a kind of rear-view mirror image of an impenetrable and opaque world that has baffled the questor. The poem is a kind of inventory of spaces and situations that are mostly beyond the range of visual identification.

Jackson Pollock. FULL FATHOM FIVE. *Collection, The Museum of Modern Art, New York, Gift of Peggy Guggenheim*

47

"To the blind all things are sudden" (Alex Leighton). There is an immediacy and unexpectedness about perception in the life of the blind. The sighted person lives habitually in a world of continuity, connectedness and perspective. He takes for granted that he can have a point of view by the mere act of arresting his movements and holding a single position, sometimes even referred to as "the correct position." By contrast the world of touch, whether passive or active, creates a relation not of connectedness but of interval. Again, if the world of the sighted naturally lends itself to a lineal connection and interrelation, the world of touch dispenses with story lines as much as melodic lines. Indeed, the visual or sighted person often entertains the idea that touch is the very essence of connectedness so that his encounter with matter tends to be as abrupt as any interval can be. Rationality itself often has the character of the merely connected and continuous, so that an untoward event, because it doesn't fit into an expected pattern, creates an involuntary arrest of tension, which is, as it were, irrational.

In cultures that give much less stress to the visual sense, "rational" connectedness exercises much less authority. Paradox and symbolic juxtaposition are accepted as natural expression. One need go no further than Japanese flower arrangement, which is achieved by means of the spaces between the flowers. In an electronic world where all-at-onceness is inevitable and normal, we have rediscovered an affinity for the discontinuity of Oriental art and expression which is most enthusiastically felt in the teen-age world today. Of course, Van Gogh, Gauguin and others anticipated the discontinuous electronic modes of perception by their immediate awareness of the new environment. Artists tend to have this power to probe and explore new environments even when most people are unhappy and uneasy about them. Theirs is not so much the power to foresee as the readiness to recognize that which is immediately present. Newspaper arrangement in the telegraph age displays many of the symbolic and unconnected features of the most advanced art.

In his great study, *The Russians as People,* Wright Miller investigates the Russian milieu, both inner and outer, concluding that in its stark monotony it was a kind of incubus. During months of the year the peasant slept on his earthen stove in a condition of utter physical involvement. The outer landscape was usually undisturbed by man-made artifacts. The low visual definition of the environment favored a high degree of tactile and acoustic stress. At this end of the sensory spectrum individuality is created by the interval of tactile involvement. At the other end of the sensory spectrum we encounter the familiar mode of individuality based on visual stress and fragmentary separateness. The visual sense lends itself to fragmentation and separateness for reasons quite antithetic to the monolithic and integral quality created by the tactile interval. Paradoxically, therefore, intense individuality is even more characteristic of the nonliterate population depicted by Dickens or Al Capp than it is of the consciously cultivated individuality of the highly literate.
The visual sense in high definition permits division and subdivision in various modes. Perhaps the most familiar of these patterns is classification. The world of bureaucracy is

the extreme example of visual organization of enterprise based on specialism and classification. It is a world very much threatened by the computer for the same reason that the continuous story line in the arts is incompatible with electronic speeds.

"Come into my parlor," said the computer to the specialist.

WORKS AND DAYS

Hesiod

90 "For ere this the tribes of men lived on earth
91 remote and free from ills and hard toil and heavy
92 sicknesses which bring the Fates upon men; for in
93 misery men grow old quickly. But the woman took
94 off the great lid of the jar with her hands and
95 scattered all these and her thoughts caused sorrow
96 and mischief to men. Only Hope remained there
97 in an unbreakable home within under the rim of the
98 great jar, and did not fly out at the door; for ere
99 that, the lid of the jar stopped her, by the will of
100 Aegis-holding Zeus who gathers the clouds. But
101 the other countless plagues wander among men;
102 for earth is full of evils and the sea is full. Of
103 themselves diseases come upon men continually by
104 day and by night, bringing mischief to mortals
105 silently; for wise Zeus took away speech from them.
106 So is there no way to escape the will of Zeus."

(Translated by Hugh G. Evelyn-White)

48

To say that Homer and Hesiod were "nonvisual" poets is to explain in a phrase every problem of the world of Greek scholarship since Lessing and Schliemann. The Greeks never entered pictorial or visual space. They tended to use all their senses at once. They approached the European modes of awareness by a gradual playing down of acoustic space, of kinetic space, of tactual and visceral spaces, in favor of a heightened visual organization of experience.

The change from multispaces to a single, uniform, rational space is often associated with the Euclidean breakthrough. In *Art and Geometry,* William Ivins explains that Euclid never freed himself from kinetic space. In *The Beginnings of Architecture,* Siegfried Giedion says that the Greeks no more managed to achieve the visual enclosure of space than did the Incas. The new space breakthrough was left for the Romans. Gombrich in *Art and Illusion* describes the rise of realistic representation in classical art: matching, not making, is the visual mode of apprehension. It still plagues our scientists when they try to explain "proof."

In the electronic age the visual properties of space and time and experience have become entirely negative, the acknowledged source of many distortions of reality. This total change in the climate of culture renders us highly appreciative once more of the mythic and the irrational, in Homer as much as in psychiatry.

Bruno Snell's *The Discovery of the Mind: The Greek Origins of European Thought* marshals the available conventional evidence, apropos Greek poetry, to show how the Greeks succeeded in sloughing off most of their nonvisual experience in order to "anticipate" European rationalism, as it were. The reader will find his book of use to the degree that he bypasses Snell's ignorance of the kinds of space created by our various senses.

It is the total unawareness of Homer's indifference to visual (continuous and connected) space that spawned all the efforts to "locate" Troy. The same misbegotten attitude motivates the efforts to establish "authorship" for *The Iliad* and *The Odyssey*. Until printing, it was obvious that all art, like language, had a collective origin. A glance at the verbs in this passage from *Works and Days* reveals that each is an assembly of metaphors. The notion of verbal "meaning" came with the rise of writing and classification. Preliterate man accepted speech as gesture and action, a sort of dance or mime of mental posture. Writing about the Homeric attitude to the multitude of forces that constitute any existing environment, Bruno Snell observes:

> According to his view—and there could be no other for him—a man's action or perception is determined by the divine forces (all technologies were Gods) operative in the world; it is a reaction of his physical organs to a stimulus, and this stimulus is itself grasped as a personal act. Any situation is likely to be the result of the stimuli, and the source of new stimuli in its turn.

The verb in line 90 is "lived" (ζώεσκον), whose Greek etymology includes "to take prisoner." (The meanings of the Greek words presented here are derived from their earliest etymologies.) For a Greek the condition of man as a prisoner of earth and existence, of his tribe and language, is a totally tactual and kinetic idea of wrap-around space (familiar to the driver of a Volkswagen today). The Japanese sign for prison is the same as that for nation or the tribal community.

In line 91 "remote" (νόσφιν) implies in Greek "aloof, secretly, clandestinely, unaided." That is, man was sufficient to himself, without corporate technologies. He led a kind of shepherd life, like the one that is the archetype of human perfection in

226

the Old Testament. (The Pandora box episode of lines 93-99 is the classic account of technological innovation and its attendant disruption of social and psychic order.)

In line 92, "bring" (ἔδωκαν) in Greek implies "to eat, consume, gnaw, waste." All the senses are active.

In line 93 "grow old" (καταγηράσκουσιν) implies in Greek "to take out the kernel." Only the shell or husk remains. This degutting process is not primarily visual or descriptive, but it is very involving.

In lines 93-94, "took off" (ἀφελοῦσα) implies in Greek to drain, suck out, a complete evisceration or degutting of the integral mode. Somewhat akin to the crash of Humpty-Dumpty.

In line 95 "scattered" (ἐσκέδασ') implies in Greek "swarming, flocking, streaming." All are implications of the rapid change of pattern that accompanies innovation and invention.

In line 95-96 "caused sorrow and mischief" (δ'ἐμήσατο) implies in the Greek "to vomit, to throw up." Technological change brings nausea and the blues in any age.

Also in line 96 "remained" (ἔμιμνε) implies "to build." The innovations demand total reconstruction of psyche and society.

In line 97 "under the rim" (ὑπὸ χείλεσιν) in Greek implies "to wet the lips, a cautious sip, a mere parting of the lips." The lesson has been learned, even if too late.

The prisoner Hope in line 98 did not "fly out" (ἐξέπτη) at the door. "Fly out" implies in Greek "attack, assault." All the innovations that had flown from the jar were onslaughts on human order. Even if the metaphor is military, it involves all the senses and spaces of sound, sight and movement.

The Greek word for "stopped her" in line 99 (ἐπέλλαβε) implies "to understand, to strengthen, to know thoroughly." By now it should be apparent that the English "translation" of these lines is quite ludicrous so far as achieving the experience of the Greek is concerned. The mirror of the mind, "held" by Zeus, dazzled and arrested the technological idiocy of the jar-opener.

In line 100, "gathers the clouds" (νεφεληγερέταο) implies in Greek a disease of the eyes. Here it is born of the divine epiphany or lightning that brings illumination to the jar-opener. (To all native societies thunder means "revelation of the divine.")

In line 101 "wander" (ἀλάληται) in Greek implies "to roam about in anguish, to be distraught by grief," as in line 100. Lines 101-105 develop the perception of the violence and misery occasioned by technological innovations.

"Come upon" (φοιτῶσι) in line 103 implies "to level, to erode, to crush." Such is the effect of technical innovation, both in Homer's time and today.

In line 104 "bringing" (φέρουσαι) implies "bearing a load, to bear a burden," but also "to lead or guide, to be carried away, swept away" as by the overwhelming force of hurricane and waves. The desperate effort to find a path through the labyrinth is swamped by the new passions unleashed by the innovation.

In line 106 "to escape" (ἐξαλέασθαι) implies "to transform utterly, to leap from the socket, or to abandon the fixed usages of tradition, to blot out."

This group of metaphors suffices to indicate the entirely nonliterary quality of Homeric language—many things are said at once, as in the puns of *Finnegans Wake*. Hesiod, Homer and Joyce inhabit a multisensuous, multispatial world "where the hand of man never set foot."

Fay Zetlin. I AM THAT I AM. *Permission of the artist*

Here a relationship between spaces that are not the spaces of pictorial space. **48**

A relationship built on the ability of the human brain to differentiate between the advancing and receding qualities of color, to live within the spatial quality generated by shape.

It could almost be defined as a subliminal awareness of the microscopic extrapolated into the macroscopic.

THE WASTE LAND

T. S. Eliot

April is the cruellest month, breeding
Lilacs out of the dead land, mixing
Memory and desire, stirring
Dull roots with spring rain.
Winter kept us warm, covering
Earth in forgetful snow, feeding
A little life with dried tubers.
Summer surprised us, coming over the Starnbergersee
With a shower of rain; we stopped in the colonnade,
And went on in sunlight, into the Hofgarten,
And drank coffee, and talked for an hour.
Bin gar keine Russin, stamm' aus Litauen, echt deutsch.
And when we were children, staying at the archduke's,
My cousin's, he took me out on a sled,
And I was frightened. He said, Marie,
Marie, hold on tight. And down we went.
In the mountains, there you feel free.
I read, much of the night, and go south in the winter.

49

Of the several kinds of space in the first eighteen lines of *The Waste Land,* the dominant space is the peculiar aloneness and isolation created by pain itself. Neurologists and biologists are quite ready to admit that pain is a mystery, but they agree that it originates in cerebral rather than sensory areas. Hence, the theoretical possibility of experiencing pain in amputated limbs or in parts of the body where there is no cause for pain, so-called "referred pain."

The world of 1922 knew much of "referred pain" and hallucinated anguish. Millions had died in 1914-18.

> "There died a myriad,
> And of the best, among them,
> For an old bitch gone in the teeth,
> For a botched civilization,
>
> Charm, smiling at the good mouth,
> Quick eyes gone under earth's lid,
>
> For two gross of broken statues,
> For a few thousand battered books."

So wrote Pound, *il miglior fabbro,* in "Hugh Selwyn Mauberley." The first two lines of *The Waste Land* compress the historical world anguish along with the seasonal or temporal misery of the conflict between winter deadness and spring freshness. Eliot alludes to the Prologue to *The Canterbury Tales* here, but unlike Eliot Chaucer had no conflict:

> "When that Aprille with hise shoures soote
> The droghte of March hath perced to the roote. . . ."

It is the beauty of April, lilac life encountering winter deadness, that creates the horror. Christ's agony of blood and tears was intensified by the beauty of the garden. (Compare

William Hogarth. IN BEDLAM (SCENE VIII FROM THE RAKE'S PROGRESS). *The Metropolitan Museum of Art, New York, Harris Brisbane Dick Fund, 1932*

Eliot's later association of Christ and spring: "In the juvescence of the year/Came Christ the tiger." Gerontion the little old man is "dead" and withered. The divine "juvescence" is not for him or his generation.) Eliot explores this paradox of pain as the space of deadness and aloneness in much of his poetry. In the first eighteen lines of *The Waste Land* he sees it both inside and out: "memory and desire," the rain and the beer garden, the mosaic of the totally unconnected events of the sleigh ride and the whodunit as nightly "distraction from distraction," the desperate flight that is contrary to nature: "I read, much of the night, and go south in the winter." The poem presents a disconnected space, psychically and socially, of anguished Peter Pans. However, the inclusiveness of the poem and the reader's deep involvement in it are made possible only by means of the mosaic pattern. Eliot has relinquished conventional visual and rational space in favor of the braillelike world of touch and the auditory spaces of resonating allusion.

49

Didn't Thomas Gray write of "Moody madness laughing wild amidst severest woe"? Hogarth has painted such a ghastly scene here. As in *The Waste Land,* pain amidst finery and romantic memories creates a conflict of emotions that defines the same space that of the pain of madness and death. It is as if one were sending roses to the Belsen or Auschwitz victims who were awaiting their turn at the gas chambers.

Beauty juxtaposed, but not connected, with cruelty or suffering is the formula for horror and madness. Anybody in a state of pain finds beauty of aspect or sound an intolerable experience. Pain creates the discontinuous spaces that we see in Bosch. Violent technical innovation creates alienation and the pain of isolation in any age. Sudden shifts of environmental stress—new sounds, greater mobility, new

sensory involvement of touch as with TV—these are experiences of extreme but unlocalizable pain for a population that has grown up in earlier spaces. Our own nostalgia for the frontier, as in *Bonanza,* is merely a mark of our alienation from new urban spaces.
Hogarth, who usually provides a narrative pattern as enclosed and richly connected as Smollett or Dickens, is in this madhouse transported beyond himself into mosaic and symbolic space by the theme of suffering and madness. There is no connection between the scenes in his canvas. It is a world of darkness. The setting is Bedlam, the madhouse which the public could visit for entertainment in the Victorian age. A Victorian commentator on Scene VIII of *The Rake's Progress* notes:

In a wretched cell, sitting naked upon the damp straw, is a despairing wretch whom fanaticism has driven mad, and who shrinks in affright from the very cross he worships. Tracing lines upon the wall is an enthusiast who believes that he has discovered longitude, and who, not being able to convince the government of the accuracy of his scheme, has gone melancholy mad;—and near him is another disappointed notary of science—an astronomer who revels in such wondrous dreams relative to the orbs that circle in the eternal realms of space that his intellect has become crushed beneath the mighty superstructure of his wild imaginings. . . . Staring in astonishment upon the astronomer is a poor tailor whose losses through the dishonesty of his aristocratic customers have driven mad. . . . On the staircase sits a poor creature crazed by love, and now absorbed in melancholy pensiveness. Behind him appears another individual . . . who believes himself to be the Pope; while in a cell opposite, a maniac simulates the stern demeanor and the imposing air which he believes consistent with the regal state whereby his disordered fancy has invested him. In the midst of the strange scene is a mad musician . . . and passing through this den of horrors are two young females whom curiosity has brought as visitors to the place. . . .

The discontinuities of space and moods in this plate lend it a kind of *avant-garde* symbolist quality.

THE EMPEROR'S NEW CLOTHES

In his poem "Esthétique du Mal" Wallace Stevens writes:

> This is the thesis scrivened in delight,
> The reverberating psalm, the right chorale.
>
> One might have thought of sight, but who could
>   think
> Of what it sees, for all the ill it sees?
> Speech found the ear, for all the evil sound,
> But the dark italics it could not propound.
> And out of what one sees and hears and out
> Of what one feels, who could have thought to make
> So many selves, so many sensuous worlds,
> As if the air, the mid-day air, was swarming

With the metaphysical changes that occur,
Merely in living as and where we live.

He indicates that the slightest shift in the level of visual intensity produces a subtle modulation in our sense of ourselves, both private and corporate. Since technologies are extensions of our own physiology, they result in new programs of an environmental kind. Such pervasive experiences as those deriving from the encounter with environments almost inevitably escape perception. When two or more environments encounter one another by direct interface, they tend to manifest their distinctive qualities. Comparison and contrast have always been a means of sharpening perception in the arts as well as in general experience. Indeed, it is upon this pattern that all the structures of art have been reared. Any artistic endeavor includes the preparing of an environment for human attention. A poem or a painting is in every sense a teaching machine for the training of perception and judgment. The artist is a person who is especially aware of the challenge and dangers of new environments presented to human sensibility. Whereas the ordinary person seeks security by numbing his perceptions against the impact of new experience, the artist delights in this novelty and instinctively creates situations that both reveal it and compensate for it. The artist studies the distortion of sensory life produced by new environmental programming and tends to create artistic situations that correct the sensory bias and derangement brought about by the new form. In social terms the artist can be regarded as a navigator who gives adequate compass bearings in spite of magnetic deflection of the needle by the changing play of forces. So understood, the artist is not a peddler of ideals or lofty experiences. He is rather the indispensable aid to action and reflection alike.

Therefore the question of whether art should be taught in our schools can easily be answered: of course it should be taught, but not as a subject. To teach art as a subject is to insure that it will exist in a state of classification serving only to separate art off from the other activities of man. As Adolf

Hildebrand points out in *The Problem of Form,* "Deflected thus from his natural course, the child develops his artificial rather than his natural resources and it is only when he reaches full maturity that the artist learns to think again in terms of the natural forces and ideas which in his childhood were his happiest possession." In the space age of information environments, art necessarily takes on new meaning and new functions. All previous classifications of these matters lose their interest and relevance.

In his *Approach to Art* E. H. Gombrich notes the extraordinary shift from making to matching that began for Western art in fifth-century Athens. In discovering the joys of matching or of realistic representation, the Greeks were not behaving like free men, but like robots. In the representation of reality stress is laid upon the visual sense usually at the expense of all the other senses. Such representation began with the rise of phonetic literacy and cannot occur at any time or at any place without the presence of a technology that favors the visual sense at the expense of all the other senses. For many people it is one of the horrors of our present age that we must live amidst the effects of technologies that do not favor the visual sense in anything like the degree that phonetic literacy does. The phonetic alphabet, as explained in *The Gutenberg Galaxy,* is the only form of writing that abstracts sight and sound from meaning. This fact is stressed by David Diringer in *The Alphabet.* By contrast, pictographic writing tends to unite the senses and semantics in a kind of gestalt. When the visual sense is played up above the other senses, it creates a new kind of space and order that we often call "rational" or pictorial space and form. Only the visual sense has the properties of continuity, uniformity and connectedness that are assumed in Euclidean space. Only the visual sense can create the impression of a continuum. Alex Leighton has said, "To the blind all things are sudden." To touch and hearing each moment is unique, but to the sense of sight the world is uniform and continuous and connected. These are the properties of pictorial space which we often confuse with rationality itself.

Perhaps the most precious possession of man is his abiding awareness of the analogy of proper proportionality, the key to all metaphysical insight and perhaps the very condition of consciousness itself. This analogical awareness is constituted of a perpetual play of ratios among ratios: A is to B what C is to D, which is to say that the ratio between A and B is proportioned to the ratio between C and D, there being a ratio between these ratios as well. This lively awareness of the most exquisite delicacy depends upon there being no connection whatever between the components. If A were linked to B, or C to D, mere logic would take the place of analogical perception. Thus one of the penalties paid for literacy and a high visual culture is a strong tendency to encounter all things through a rigorous story line, as it were. Paradoxically, connected spaces and situations exclude participation whereas discontinuity affords room for involvement. Visual space is connected and creates detachment or noninvolvement. It also tends to exclude the participation of the other senses. Thus the New York World's Fair defeated itself by imposing a visual order and story line that offered little opportunity for participation by the viewer. In contrast, Expo Canada presented not a story line but a mosaic of many cultures and environments. Mosaic form is almost like an X-ray compared to pictorial form with its connections. The Canadian mosaic aroused extraordinary enthusiasm and participation, mystifying many people.

The same difference exists between movie and TV. The movie is highly pictorial, but kinematically it is discontinuous and nonvisual, and thus demands participation. This discontinuous quality has been very much played up in such movies as *The Seventh Seal* and *Blow-Up,* to name only two in a rapidly developing métier. A movie is a succession of discrete images which are separated by extremely small spans of time. Because of their rapid succession, the images are fused in the conscious mind and appear connected. Our relatively recent insights into the power of the preconscious in both the creation and the apprehension of works of art indicate that the subliminal is in fact a strong force in psychic reorganization. It is in this sense that the movie form can be described as a medium which deals in disconnected spaces.

TV, on the other hand, is a kind of X-ray. Any new technology, any extension or amplification of human faculties, when given material embodiment, tends to create a new environment. This is as true of clothing as of speech, or script, or wheel. This process is more easily observed in our own time when several new environments have been created. In the latest one, TV, we find a handful of engineers and technicians in the 10 percent area, as it were, creating a set of radical changes in the 90 percent area of daily life. The new TV environment is an electric circuit that takes as its content the earlier environment, the photograph and the movie in particular. The interplay between the old and the new environments generates an innumerable series of problems and confusions which extend all the way from how to allocate the viewing time of children and adults to pay-TV and TV in the classroom. The new medium of TV as an environment creates new occupations. As an environment, it is imperceptible except in terms of its content. That is, all that is seen or noticed is the old environment, the movie. But even the effects of TV on the movie go unnoticed, and the effects of the TV environment in altering the entire character of human sensibility and sensory ratio are completely ignored. The viewer is in the situation of being X-rayed by the image. Typically, therefore, the young viewer acquires a habit of depth involvement which alienates him from the existing arrangements of space and organized knowledge, whether at home or in the classroom. However, this condition of alienation extends to the entire situation of Western man today.

The function of the artist in correcting the unconscious bias of perception in any given culture can be betrayed if he merely repeats the bias of the culture instead of readjusting it. In fact, it can be said that any culture which feeds merely on its direct antecedents is dying. In this sense the role of art is to create the means of perception by creating counterenvironments that open the door of perception to people otherwise numbed in a nonperceivable situation. In Françoise Gilot's book *Life with*

Picasso the painter notes that: "When I paint, I always try to give an image people are not expecting and, beyond that, one they reject. That's what interests me. It's in this sense that I mean I always try to be subversive. That is, I give a man an image of himself whose elements are collected from among the usual way of seeing things in traditional painting and then reassembled in a fashion that is unexpected and disturbing enough to make it impossible for him to escape the questions it raises."

Under the heading "What exists is likely to be misallocated" Peter Drucker in *Managing for Results* discusses the structure of social situations: "Business enterprise is not a phenomenon of nature but one of society. In a social situation, however, events are not distributed according to the 'normal distribution' of a natural universe (that is, they are not distributed according to the bell-shaped Gaussian curve). In a social situation a very small number of events *at one extreme* —the first 10 per cent to 20 per cent at most—account for 90 per cent of all results." What Drucker is presenting here is the environment as it presents itself for human attention and action. He is confronting the phenomenon of the imperceptibility of the environment as such. Edward T. Hall tackles this same factor in *The Silent Language*. The ground rules, the pervasive structure, the over-all pattern elude perception except insofar as an antienvironment or a countersituation is constructed to provide a means of direct attention. Paradoxically, the 10 percent of the typical situation that Drucker designates as the area of effective cause and as the area of opportunity, this small factor, is the environment. The other 90 percent is the area of problems generated by the active power of the 10 percent environment. For the environment is an active process pervading and impinging upon all the components of the situation. It is easy to illustrate this.

The content of any system or organization naturally consists of the preceding system or organization, and in that degree the old environment acts as a control on the new. It is useful to notice that the arts and sciences serve as antienvironments that enable us to perceive the environment. In a business civilization we have long considered liberal study as providing necessary means of orientation and perception. When the arts and sciences themselves become environments under conditions of electric circuitry, conventional liberal studies, whether in the arts or sciences, will no longer serve as an antienvironment. When we live in a museum without walls, or have music as a structural part of our sensory environment, new strategies of attention and perception have to be created. When the highest scientific knowledge creates the environment of the atom bomb, new controls for the scientific environment have to be discovered, if only in the interest of survival.

The structural examples of the relation of environment to antienvironment need to be multiplied in order to understand the principles of perception and activity involved. The Balinese, who have no word for art, say, "We do everything as well as possible." This is not an ironic but a factual remark. In a preliterate society art serves as a means of merging the individual and the environment, not as a means of training perception of the environment. Archaic or primitive art looks to us like a magical control built into the environment. Thus to put the artifacts from such a culture into a museum or antienvironment is an act of nullification rather than of revelation. Today what is called "Pop Art" is the use of some object from our own daily environment as if it were antienvironmental. Pop Art serves to remind us, however, that we have fashioned for ourselves a world of artifacts and images that are intended not to train perception or awareness but to insist that we merge with them as the primitive man merges with his environment. Therefore, under the terms of our definition of art as antienvironmental, this is nonart except insofar as the illumination of the interior environment of the human mind can be regarded as an artistic stance.

The world of modern advertising is a magical environment constructed to maintain the economy, not to increase human awareness. We have designed schools as antienvironments to develop the perception and judgment of the printed word, but we have provided no training to develop similar perception and judgment of any of the new environments created by electric circuitry. This is not accidental. From the development of phonetic script until the invention of the electric telegraph, human technology had tended strongly toward the furtherance of detachment and objectivity, detribalization and individuality. Electric circuitry has quite the contrary effect. It involves in depth. It merges the individual and the mass environment. To create an antienvironment for such electric technology would seem to require a technological extension of both private and corporate consciousness. The awareness and opposition of the individual are in these circumstances as irrelevant as they are futile.

The structural features of environment and antienvironment appear in the age-old clash between professionalism and amateurism, whether in sport or in studies. Professional sport fosters the merging of the individual in the mass and in the patterns of the total environment. Amateur sport seeks rather the development of critical awareness of the individual and, most of all, critical awareness of the ground rules of the society as such. The same contrast exists for studies. The professional tends to specialize and to merge his being uncritically in the mass. The ground rules provided by the mass response of his colleagues serve as a pervasive environment of which he is uncritical and unaware.

The party system of government affords a familiar image of the relations of environment and antienvironment. The government as environment needs the opposition as antienvironment in order to be aware of itself. The role of the opposition seems to be, as in the arts and sciences, that of creating perception. As the government environment becomes more cohesively involved in a world of instant information, opposition would seem to become increasingly necessary but also intolerable. It begins to assume the rancorous and hostile character of a Dew Line, or a Distant Early Warning System. It is important, however, to consider the role of the arts and sciences as Early Warning Systems in the social environment.

The models of perception they provide can give indispensable orientation to future problems well before they become troublesome.

The legend of Humpty-Dumpty suggests a parallel to the 10-90 percent distribution of causes and effects. His fall brought into play a massive response from the social bureaucracy. But all the King's horses and all the King's men could not put Humpty-Dumpty back together again. They could not re-create the old environment; they could only create a new one. Our typical response to a disrupting new technology is to re-create the old environment instead of heeding the new opportunities of the new environment. Failure to notice the new opportunities is also failure to understand the new powers. This means that we fail to develop the necessary controls or antienvironments for the new environment. This failure leaves us in the role of mere automata.

W. T. Easterbrook has done extensive exploration of the relations of bureaucracy and enterprise, discovering that as soon as one becomes the environment, the other becomes an antienvironment. They seem to bicycle along through history alternating their roles with all the dash and vigor of Tweedledum and Tweedledee. In the eighteenth century when *realism* became a new method in literature, what happened was that the external environment was put in the place of antienvironment. The ordinary world was given the role of art object by Daniel Defoe and others. The environment began to be used as a perceptual probe. It became self-conscious. It became an "anxious object" instead of being an unperceived and pervasive pattern. Environment used as probe or art object is satirical because it draws attention to itself. The Romantic poets extended this technique to external nature, transforming nature into an art object. Beginning with Baudelaire and Rimbaud and continuing in Hopkins and Eliot and James Joyce, the poets turned their attention to language as a probe. Long used as an environment, language became an instrument of exploration and research. It became an antienvironment. It became Pop Art along with the graphic probes of Larry Rivers, Rauschenberg and many others.

The artist as a maker of antienvironments permits us to perceive that much is newly environmental and therefore most active in transforming situations. This would seem to be why the artist has in many circles in the past century been called the enemy, the criminal.

> Pablo shook his head. "Kahnweiler's right," he said. "The point is, art is something subversive. It's something that should *not* be free. Art and liberty, like the fire of Prometheus, are things one must steal, to be used against the established order. Once art becomes official and open to everyone, then it becomes the new academicism." He tossed the cablegram down onto the table. "How can I support an idea like that? If art is ever given the keys to the city, it will be because it's been so watered down, rendered so impotent, that it's not worth fighting for."
>
> I reminded him that Malherbe had said a poet is of no more use to the state than a man who spends his time playing ninepins. "Of course," Pablo said. "And why did Plato say poets should be chased out of the republic? Precisely because every poet and every artist is an antisocial being. He's not that way because he wants to be; he can't be any other way. Of *course* the state has the right to chase him away—from *its* point of view—and if he is really an artist it is in his nature not to want to be admitted, because if he is admitted it can only mean he is doing something which is understood, approved, and therefore old hat—worthless. Anything new, anything worth doing, can't be recognized. People just don't have that much vision." (Françoise Gilot and Carlton Lake, *Life with Picasso*)

It helps to explain why news has a natural bias toward crime and bad news. It is this kind of news that enables us to perceive our world. The detective since Poe's Dupin has tended to be a probe, an artist of the big town, an artist-enemy, as it were. Conventionally, society is always one phase back, is never environmental. Paradoxically, it is the antecedent environment that is always being upgraded for our attention. The new environment always uses the old environment as its material.

In the Spring, 1965, issue of the *Varsity Graduate* of the University of Toronto, Glenn Gould discussed the effects of recorded music on performance and composition. This is a reversal or chiasmus of form that occurs in any situation where an environment is pushed up into high intensity or high definition by technological change. A reversal of characteristics occurs, as in the case of bureaucracy and enterprise. An environment is naturally of low intensity or low definition. That is why it escapes observation. Anything that raises the environment to high intensity, whether it be a storm in nature or violent change resulting from a new technology, turns the environment into an object of attention. When it becomes an object of attention, it assumes the character of an antienvironment or an art object. When the social environment is stirred up to exceptional intensity by technological change and becomes a focus of much attention, we apply the terms "war" and "revolution." All the components of "war" are present in any environment whatsoever. The recognition of war depends upon their being stepped up to high definition.

Under electric conditions of instant information movement, both the concept and the reality of war become manifest in many situations of daily life. We have long been accustomed to war as that which goes on between publics or nations. Publics and nations were the creation of print technology. With electric circuitry publics and nations became the content of the new technology: "The mass audience is not a public as environment but a public as content of a new electric environment." And whereas "the public" as an environment created by print technology consisted of separate individuals with varying points of view, the mass audience consists of the same individuals involved in depth in one another and involved in the creative process of the art or educational situation that is presented to them. Art and education were presented to the *public* as consumer packages for their instruction and edification. The members of the mass audience are immediately involved in art and education as participants and co-creators rather than as consumers. Art and education become new forms of experience, new environments, rather than new antienvironments. Pre-electric art and education were antienvironments in the sense that they were the content of various environments. Under electric conditions the content tends, however, toward becoming environmental itself. This was the paradox that Malraux found in *The Museum Without Walls,* and that Glenn Gould finds in recorded music. Music in the concert hall had been an antienvironment. The same music when recorded is *music without halls,* as it were.

Another paradoxical aspect of this change is that when music becomes environmental by electric means, it becomes more and more the concern of the private individual. By the same token and complementary to the same paradox, the pre-electric music of the concert hall (the music made for a public rather than a mass audience) was a corporate ritual for the group rather than the individual. This paradox extends to all electric technology. The same means which permit a universal and centralized thermostat in effect encourage a private thermostat for individual manipulation. The age of the mass audience is thus far more individualistic than the preceding age of the *public.* It is this paradoxical dynamic that confuses every issue about "conformity," "separatism" and "integration" today. Profoundly contradictory actions and directions prevail in all these situations. This is not surprising in an age of circuitry succeeding the age of the wheel. The feedback loop plays all sorts of tricks to confound the single-plane and one-way direction of thought and action as they had been constituted in the pre-electric age of the machine.

Applying the foregoing to the Negro question, one could say that the agrarian South has long tended to regard the Negro as environment. As such, the Negro is a challenge, a threat, a burden. The very phrase "white supremacy," quite as much as the phrase "white trash," registers this environmental attitude. The environment is the enemy that must be subdued. To the rural man, the conquest of nature is an unceasing challenge. It is the Southerner who contributed the cowboy to the frontier. The Virginian, the archetypal cowboy, as it were, confronts the environment as a hostile, natural force. To man on the frontier, other men are environmental and hostile. By contrast, to the townsmen, men appear not as environmental but as content of the urban environment.

Parallel to the Negro question is the problem of French Canada. The English Canadians have been the environment of French Canada since the railway and Confederation. However, since the telegraph and radio and television, French Canada and English Canada alike have become the content of this new technology. Electric technology is totally environmental for all human communities today. Hence the great confusion arising from the transformation of environments into antienvironments, as it were. All the earlier groupings that had constituted separate environments before electricity have now become antienvironments or the content of the new technology. Awareness of the old unconscious environments therefore becomes increasingly acute. The content of any new environment is just as unperceived as that of the old one had been initially. As a merely automatic sequence, the succession of environments and the dramatics accompanying them tend to be rather tiresome, if only because the audience is very prone to participate in the dramatics with an enthusiasm proportionate to its lack of awareness. In the electric age all former environments whatever become antienvironments. As such the old environments are transformed into areas of self-awareness and self-assertion, guaranteeing a very lively interplay of forces.

The visual sense, alone of our senses, creates the forms of space and time that are uniform, continuous and connected. Euclidean space is the prerogative of visual and literate man. With the advent of electric circuitry and the instant movement of information, Euclidean space recedes and the non-Euclidean geometries emerge. Lewis Carroll, the Oxford mathematician, was perfectly aware of this change in our world when he took Alice through the looking glass into the world where each object creates its own space and conditions. To the visual or Euclidean man, objects do not create time and space. They are merely fitted into time and space. The idea of the world as an environment that is more or less fixed is very much the product of literacy and visual assumptions. In his book *The Philosophical Impact of Contemporary Physics* Milic Capek explains some of the strange confusions in the scientific mind that result from the encounter of the old non-Euclidean spaces of preliterate man with the Euclidean and Newtonian spaces of literate man. The scientists of our time are just as confused as the philosophers, or the teachers, and it is for the reason that Whitehead assigned: they still have the illusion that the new developments are to be fitted into the old space or environment.

One of the most obvious changes in the arts of our time has been the dropping not only of representation, but also of the story line. In poetry, in the novel, in the movie, narrative continuity has yielded to thematic variation. Such variation in place of story line or melodic line has always been the norm in native societies. It is now becoming the norm in our own society and for the same reason, namely, that we are becoming a nonvisual society.

In the age of circuitry, or feedback, fragmentation and specialism tend to yield to integral forms of organization. Humpty-Dumpty tends to go back together again. The bureaucratic efforts of all the King's horses and all the King's men were naturally calculated to keep Humpty-Dumpty from ever getting together again. The Neolithic age, the age of the planter after the age of the hunter, was an age of specialism and division of labor. It has reached a somewhat startling terminus with the advent of electric circuitry. Circuitry is a profoundly decentralizing process. Paradoxically, it was the wheel and mechanical innovation that created centralism. The circuit reverses the characteristics of the wheel, just as Xerography reverses the characteristics of the printing press. Before printing, the scribe, the author and the reader tended to merge. With printing, author and publisher became highly specialized and centralized forms of action. With Xerography, author and publisher and reader tend to merge once more.

Whereas the printed book had been the first mass-produced product, creating uniform prices and markets, Xerography tends to restore the custom-made book. Writing and publishing tend to become services of a corporate and inclusive kind. The printed word created the Public. The Public consists of separate individuals, each with his own point of view. Electric circuitry does not create a Public. It creates the Mass. The Mass does not consist of separate individuals, but of individuals profoundly involved in one another. This involvement is a function not of numbers but of speed.

The daily newspaper is an interesting example of this fact. The items in the daily press are totally discontinuous and totally unconnected. The only unifying feature of the press is the date line. Through that date line the reader must go, as Alice went, "through the looking glass." If it is not today's date line, he cannot get in. Once he goes through the date line, he is involved in a world of items for which he, the reader, must write a story line. He makes the news, as the reader of a detective story makes the plot. In the same way the relatively open-ended movie at the Czech pavilion in Expo allowed for intense audience participation through the easy availability of the consensus.

Just as the printing press created the Public as a new environment, so does each new technology or extension of our physical powers tend to create new environments. In the age of information, it is information itself that becomes environmental. The satellites and antennae projected from our planet, for example, have transformed the planet from being an environment into being a probe. This is a transformation which the artists of the past century have been explaining to us in their endless experimental models. Modern art, whether in painting or poetry or music, began as a probe and not as a package. The Symbolists literally broke up the old packages and put them into our hands as probes. And whereas the package belongs to a consumer age, the probe belongs to an age of experimenters.

One of the peculiarities of art is to serve as an antienvironment, a probe that makes the environment visible. It is a form of symbolic, or parabolic, action. Parable comes from a word that means literally "to throw against," just as symbol comes from one meaning "to throw together." As we equip the planet with satellites and antennae, we tend to create new environments of which the planet is itself the content. It is peculiar to environments that they are complex processes which transform their content into archetypal forms. As the planet becomes the content of a new information environment, it also tends to become a work of art. Where railway and machine created a new environment for agrarian man, the old agrarian world became an art form. Nature became a work of art. The Romantic movement was born. When the electric circuit enveloped the mechanical environment, the machine itself became a work of art. Abstract art was born.

As information becomes our environment, it becomes mandatory to program the environment itself as a work of art. The parallel to this appears in Jacques Ellul's *Propaganda,* where he sees propaganda not as an ideology or content of any medium, but as the operation of all the media at once. The mother tongue is propaganda because it exercises an effect on all the senses at once. It shapes our entire outlook and all our ways of feeling. Like any other environment, its operation is imperceptible. When an environment is new, we perceive the old one for the first time. What we see on the Late Show is not TV, but old movies. When the Emperor appeared in his new clothes, his courtiers did not see his nudity, they saw his old clothes. Only the small child and the artist have the immediacy of approach that permits perception of the environmental. The artist provides us with antienvironments that enable us to see the environment. Such antienvironmental means of perception must constantly be renewed in order to be efficacious. That basic aspect of the human condition by which we are rendered incapable of perceiving the environment is one to which psychologists have not even referred. In an age of accelerated change, the need to perceive the environment becomes urgent. Acceleration also makes such perception of the environment more possible. Was it not Bertrand Russell who said that if the bath water got only half a degree warmer every hour, we would never know when to scream? New environments reset our sensory thresholds. These, in turn, alter our outlook and expectations.

The need of our time is for a means of measuring sensory thresholds and a means of discovering exactly what changes occur in these thresholds as a result of the advent of any particular technology. With such knowledge in hand, it would be possible to program a reasonable and orderly future for any human community. Such knowledge would be the equivalent of a thermostatic control for room temperatures. It would seem only reasonable to extend such controls to all the sensory thresholds of our being. We have no reason to be grateful to those who haphazardly juggle the thresholds in the name of innovation.

Redesign of the so-called "light shows" so that they cease to be merely bombardment and become probes into the environment would be most beneficial in an educational sense.

The Two Cultures by C. P. Snow is a handy instance of our contemporary dilemma between visual and nonvisual methods of codifying and processing reality (C. P. Snow seems to be blowing both horns of the dilemma). The dilemma is the same as that which confronted Alice in *Through the Looking Glass*. Before she went through the looking glass, she was in a visual world of continuity and connected space where the appearance of things matched the reality. When she went through the looking glass, she found herself in a nonvisual world where nothing matched and everything seemed to have been made on a unique pattern. (As a matter of fact, because of electric technology we do have two cultures. They are the culture of our children and that of ourselves; we don't dialogue.) The work of Robert Ardrey in *The Territorial Imperative* is a kind of report from Alice after she had gone through the looking glass. Territoriality is the power of things to impose their own assumptions of time and space by means of our sensory involvement in them. Again, it is a world of making rather than of matching. Modern physics in general carries us into an unvisualizable territory. The speeds as well as the submicroscopic character of its particles are beyond visual representation. John R. Platt in *The Step to Man* explains how it would be possible to incorporate the twenty million books in the world today into an electronic library no larger than the head of a pin.

The present concern with "the death of God" is very much related to the decline in visual culture. The theologian Altizer tells us that the death of God happened roughly two hundred years ago "when the understanding of history grew to supplant an old God-concept. The Christ preserved by the Church has been so progressively dissolved and the God it preached so far decomposed that it is not possible to begin to see Jesus as the core of faith and as incarnate in humanity wherever there is life, and to see God as the opposite of humanity, life, progress—that is, as death." (James Heisig, *The Wake of God,* Divine Seminary, 1967.) In a visual sense God is no longer "up there" and "out there" any more than twenty million books in a pinhead could be said to be "in there." Visual orientation has simply become irrelevant. Some feel that Christianity's existence must always stand in the tension between being in the world and standing outside it. Kierkegaard was keenly aware of this, as were St. Paul and, later, Martin Luther. But the tension between inner and outer is a merely visual guideline, and in the age of the X-ray inner and outer are simultaneous events.

As the Western world goes Oriental on its inner trip with electric circuitry, it is not only the conventional image of God that is deposed; the whole nature of self-identity enters a state of crisis. God the clockmaker and engineer of the universe is no more an essential visual image to the West than is the identity card or the visual classification as an image of private personal status. The problem of personal identity first arose in the West with King Oedipus, who went through the crisis of detribalization, the loss of corporate involvement in the tribal group. To an ancient Greek the discovery of private identity was a terrifying and horrible thing that came about with the discovery of visual space and fragmentary classification. Twentieth-century man is traveling the reverse course, from an extreme individual fragmentary state back into a condition of corporate involvement with all mankind. Paradoxically, this new involvement is experienced as alienation and loss of private selfhood. It began with Ibsen and the Russian writers like Dostoyevsky, for whom there remained a much larger degree of awareness of the old tribal and corporate life than anything available to other European writers in the nineteenth century. The novelists and dramatists who began the quest to discover "Who am I?" have been succeeded by the existentialist philosophers, who meditate upon the meaninglessness of private lives in the contemporary world:

> One can say, in short, that meaninglessness is spreading before our eyes. A strange inner mutation is thereby produced which takes on the aspect of a genuine uprooting. Entirely new questions are being asked, they insist upon being asked, where one hitherto seemed to be in an order which contained its own justification; it is the very order to which the barracks man belonged in the days when he was still a living being, when he was in the present.
> He for whom reflection has become a need, a primordial necessity, becomes aware of the precarious and contingent character of the conditions which constitute the very framework of his existence. The word "normal," which he once made use of in a way which now seems to him so imprudent, is emptied of its significance—let us say at least that it is suddenly, as it were, marked by a sign which makes it appear in a new and disturbing light. (Gabriel Marcel, *Problematic Man,* Herder and Herder, 1967)

Marcel is quite aware that there are no concepts or categories that can resolve this crisis:

> Let us now go back to the questions which the barracks man was asking himself: *Who am I? What sense does my life have?* It is obvious that one does not resolve these questions by saying to this

man (or to myself if I ask them of myself): You are a rational animal. An answer of this kind is beside the point. I said earlier that meaninglessness was spreading: that is to say that I, who have a profession, a country, means of existence, etc., cannot help but turn these questions somehow towards myself. Why is this so? Let us reason *a contrario,* and suppose that I shut myself up prudently, jealously, in that favored category where these questions do not arise. But if I have really managed, by an effort of imagination, to put myself in the place of the barracks man, it is through his eyes that I will be brought to consider the step by which I placed myself once and for all in the category of the privileged, who know who they are, and what they are living for. In other words, by the combined action of imagination and reflection, I have been able to bring about a change which bears not only upon the object, but upon the subject himself, the subject who questions.

However, he seems to favor the illusion that these dilemmas are ideological in origin rather than a consequence of a reprogramming of the human environment in its sensory modes. The rear-view mirror is the favorite instrument of the philosophical historian:

> In particular, one can hardly contest the fact that nationalism in its modern, post-revolutionary form is the product of an ideology that developed in the eighteenth century and combined, under conditions very difficult to state precisely, with a pre-romanticism whose origins seem to be found in Rousseau. Abandoned to its own inclination, this ideology led to a kind of cosmopolitanism of reason. The nationalism which issued from the French Revolution built itself to a large extent upon the ruins of the basic communities which had persisted until the end of the *ancien régime,* but which the individualism of the philosophy of the Enlightenment inevitably helped to dissolve. One cannot deny, on the other hand, that there was a close connection between this fact and the devitalization of religion which occurred in the same period. But the industrial revolution, at least during the first part of the nineteenth century, was destined to play a part in considerably aggravating this tendency—to a large extent, moreover, under the influence of a liberalism which on the economic plane (as we know all too well) was destined to engender the most inhuman consequences, the individual being reduced to a more and more fragmentary condition, under the cover of an optimism which seems to us today to have been the height of hypocrisy.

Marcel occasionally entertains the possibility of considering existence not as a classification or category, but as a total environment:

> The profound justification of the philosophies of existence has perhaps consisted above all in the fact that they have brought out the impossibility of considering an existent being without taking into consideration his existence, his mode of existence. But regarding this very existence, the words *rational animal* furnish us no genuine enlightenment.

But in general he is aware of the futility of history. In the electric age, however, history no longer presents itself as a perspective of continuous visual space, but as an all-at-once and simultaneous presence of all facets of the past. This is what T. S. Eliot calls "tradition" in his celebrated essay "Tradition and the Individual Talent." Eliot's concept seemed quite revolutionary in 1917, but it was in fact a report of an immediate and present reality. Awareness of all-at-once history or tradition goes with a correlative awareness of the present as modifying the entire past. It is this vision that is characteristic of the artistic perception which is necessarily concerned with making and change rather than with any point of view or any static position.

The bourgeois nineteenth century referred only to those faces and features which were most strikingly visual in their tidiness and order. That world now persists in some degree in suburbia with the Educational Establishment as its sustaining bulwark. Antithetic to suburbia is the beatnik world, which in the nineteenth century was Bohemia. This is a world in which visual values play a very minor role. One hippie was heard to say, "I have no use for this Cromwell character. I'm a Cavalier!" Cromwell was a sort of *avant-garde* program of visual values. His "Ironsides" were an advance image of industrial production and weaponry. Their "Roundheads" are now the "square" citizens of the upper executive world. "Square," of course, simply means visual and uninvolved.

The transition between worlds may have occurred at the

moment of the hula hoop. Mysteriously, people were fascinated by hula hoops as an invitation to involvement and gyration, but nobody was ever seen rolling one in the approved style of the hoop and stick of yesteryear. When exhorted by their elders to roll these hoops down a walk, children simply ignored the request. An equivalent situation today is the disappearance of the word "escapism" in favor of the word "involvement." In the twenties all popular art, whether written or photographed for the movies, was branded as pure escapism. It has not occurred to anybody to call TV viewing escapist any more than it had occurred to anybody to roll the hula hoop as though it were a wheel. Today popular art is intensely involving, and it contains none of the visual values that characterized respectable art a century ago. Popular art has indeed swamped Bohemia and enlarged its territories many times. The aesthete, 1967 model, does not affect any nineteenth-century elegance, but in the interest of involvement presents a shaggy and multisensuous image. Upon meeting him we may well be inclined to say, "You're putting me on!" This is indeed the case. The image to which both beatnik and Beatle aspire is that of "putting on" the corporate audience. It is not a private need of expression that motivates them, but a corporate need of involvement in the total audience. This is humanism in reverse, instead of the corporate image of an integral society.

The revolt against the exclusively humanistic conception of art has been long in gestation, but it first comes into visible existence in the painting of Cézanne, and Cézanne's fundamental importance in the history of this revolution is due precisely to the fact that he was the first who dared assert that the purpose of art is not to express an ideal, whether religious or moral or humanistic, but simply to be humble before nature, and to render the forms which close observation could disentangle from vague visual impressions. The consequences of this peculiar kind of honesty were hardly such as Cézanne himself would have expected. First came cubism, and then a gradual purification of form which reached its logical conclusion in the abstract or nonfigurative art of Piet Mondrian or Ben Nicholson. This formalist type of art is now widespread among artists in every medium, and whether you like it or not, like technology it has come to stay. (Herbert Read, *The Redemption of the Robot*, Trident Press, 1966)

A somewhat different approach to the problem of the transforming action of new environments upon older ones can be taken by the study of cliché and archetype. The world of the cliché is itself environmental since nothing can become a cliché until it has pervaded some world or other. It is at the moment of pervasiveness that the cliché becomes invisible. In their study of *The Popular Arts* Stuart Hall and Paddy Whannel have provided many illustrations of the principle by which a world of cliché, by the art of enveloping an older cliché, seems to turn the older cliché into an archetype or art form. They point to the world of Mickey Spillane, in which the free-lance avenger saves the law by working outside it. Raymond Chandler is much more sophisticated:

> As Chandler's work develops, his themes emerge with greater clarity. When he died he was still at work on *The Poodle Springs Story*. This was to be only incidentally a thriller. Marlowe, married to a wealthy girl, is in danger of becoming her "poodle," confined to the empty round of California cocktail parties. "The contest between what she wants Marlowe to do and what he will insist on doing will make a good sub-plot. I don't know how it will turn out, but she'll never tame him. Perhaps the marriage won't last, or she might even learn to respect his integrity," Chandler wrote, ". . . a struggle of personalities and ideas of life": the thriller becoming the novel of manners.

It is not only that a new medium creates a new environment, which acts upon the sensory life of its inhabitants. The same new environment acts upon the older literary and artistic forms as well:

> As these various satirical modes are more fully employed we begin to understand Chandler's real achievement. Like the true satirist, his gift lies in a disenchanted view of life, and depends upon a highly artificial style. Like the mock-heroic writers and poets, who made play with "heroism," Chandler makes play with

the notion of "toughness." He inverts the thriller conventions, draws attention to their artificiality. A hard, polished prose surface permits his wit to play freely. Where the lesser practitioners in the field break their necks to build up the arch-hero, the superman, at the centre of their work, Chandler sets out to portray the most practised of anti-heroes. Apart from Marlowe, who is keeper of both conscience and consciousness in the novel, and through whose elliptical eye every detail is observed and placed, few of the other characters have true "depth." They are consciously two-dimensional, like the characters in a Ben Jonson play or in Restoration comedy. Perhaps, like the latter, a Chandler novel is a decadent work of art, and there are signs of this in the language (for one thing the similes tend to be over-elaborate and ornate or bizarre). But his use of the witticism or the wisecrack has the same pointed "surface" effect as the rhymed couplet or the epigram in Restoration comedy. There are countless effects of a literary kind which lesser novelists, practising in the more major literary genres, are able to achieve, but which escaped Chandler. But there are many compensating pleasures which are not to be found in their work. Few writers have used so compromised and over-worked a popular literary form with such skill, craftsmanship and tact.

The hero of the modern thriller puts on the audience, as it were, in a typical gesture of total involvement, whereas the hero of the older adventure story was an aristocratic individual. The new hero is a corporate rather than a private individual figure:

As Orwell showed in his comparison between *Raffles* and *Miss Blandish*, the modern thriller-hero can no longer afford to stand aside from the action in his story with that aristocratic detachment which was possible in his immediate predecessors. Unlike Sherlock Holmes or Lord Peter Wimsey or that meticulous *deus ex machina* Hercule Poirot, the thriller-hero must finally enter the action as the main protagonist. The omniscience of the earlier detective-heroes provided some distance between them and the mere mortals caught up in the drama and confusion of the crime. But now this hero, of all the figures in the novel, must be the *most* exposed to the play of passion and violence, the one most intimately caught up with the actual experience of punishment. And if we ask why this change has come about we are forced to give a complex set of reasons, all of which suggest how deeply rooted the literature is in the social imagination. Perhaps it is because we can no longer accept the figure who stands outside the action and yet knows all the answers: we demand greater verisimilitude today. Perhaps it is because these impersonal figures seem now too superhumanly remote: since the revolution in our thinking effected by Freud and psychoanalysis, we take a different view of crime, punishment and violence which the thriller reflects. We cannot believe in the hero who is himself wholly free from the inner compulsions of violence and lawlessness —we demand that he should stand closer to the villain, exposed to the very evils he is dedicated to remove: "there, but for the grace of God . . ." Certainly, the philosopher would argue that the thriller also shows a collapse in the belief in an abstract and incorruptible justice.

What Hall and Whannel are saying is that the new hero is constituted differently by virtue of being representative of the entire reading public.

The Mike Hammer and James Bond stories are, of course, fantasies—but fantasies which communicate a graphic and heightened realism. Characters may be overdrawn, situations stereo-typed, resolutions predictable. But the fictional life of these stories is convincing at the very level at which the modern reader, especially the young reader, is likely to find himself most under pressure: at the level of the sensations. In a quite precise sense, the thriller novel is a novel of the sensations. Its power lies in its experiential quality, in the absence of relieving factors and the starkness of the action, and in the image of human behaviour which it offers.

In exactly the same way the modern painting does not allow for the single point of view or the dispassionate survey. The modern painter offers an opportunity for dialogue within the parameters inherent in an art form which is moving away from the rational-visual and into the total world of man's sensory involvement.

APPENDICES

I. A NOTE ON TACTILITY

In a visual culture it sounds quite paradoxical to say that sculpture is primarily tactile and only incidentally visual. In fact, tactility is a matter that has scarcely been discussed, and yet it is crucial in the world of the arts. There are occasional passages on the subject, like Ortega y Gasset's in *Man and People* (Norton):

   And if it is true that the visible and the act of seeing afford greater clarity as examples in a first approach to our doctrine, it would be a grave mistake to suppose that sight is the most important "sense." Even from the psycho-physiological point of view, which is ancillary, it seems more and more probable that touch was the original sense from which the others were gradually differentiated. From our more radical point of view it is clear that the decisive form of our intercourse with things is in fact touch. And if this is so, touch and contact are necessarily the most conclusive factor in determining the structure of our world.

Yet Ortega y Gasset scarcely develops this theme and when he does illustrate it, it is without any reference whatever to tactility. For example, when he is discussing Herbert Spencer and his doctrine of the handshake, he is in the very center of tactility. Tactility is the world of the interval, not of the connection, and that is why it is antithetic to the visual world. For the visual is above all the world of the continuous and the connected. For visual or civilized man, the handshake has reversed its original function and meaning of stressing the cultural interval or division between peoples. An analogue to the handshake as tactile and as stressing interval exists in the Oriental world, where, as Ortega y Gasset writes:

Courtesy, as we shall later see, is a social technique that eases the collision and strife and friction that sociality is. Around each individual it creates a series of tiny buffers that lessen the other's bump against us and ours against the other. The best proof that this is so lies in the fact that courtesy was able to attain its most perfect, richest, and most refined forms in countries whose population density was very great. Hence, it reached its maximum where that is highest—namely, in the Far East, in China and Japan, where men have to live too close to one another, almost on top of one another. Without all those little buffers, living together would be impossible. It is well known that the European in China produces the impression of a rude, crass, and thoroughly ill-educated being. So it is not surprising that the Japanese language has succeeded in suppressing those two slightly and sometimes more than slightly impertinent pistol-shots, the *you* and the *I*, in which, whether I want to or not, I inject my personality into my neighbor and my idea of his personality into the *You*.

The strategy of courtesy as a means of maintaining social interval is of the very essence of tactility.

In music, however, it receives even more notable stress. Both upbeat and downbeat are tactile modes that create the separation whose closure makes a unique rhythm. The matter is at least alluded to in Harold Schonberg's *The Great Conductors* (Simon & Schuster), when he cites the observation of Richard Strauss:

> In fifty years of practice, I have discovered how unimportant it is to beat each quarter note or eighth note. What is decisive is that the upbeat, which contains the whole of the tempo that follows, should be rhythmically exact and that the downbeat should be extremely precise. The second half of the bar is immaterial. I frequently conduct it like an *alla breve* (i.e., in twos). Always conduct in periods, never bars. . . . Second rate conductors are frequently inclined to pay too much attention to the elaborations of rhythmic detail, thus overlooking the proper shaping of a phrase as a whole. . . . Any modification of tempo made necessary by the character of the piece should be carried out imperceptibly, so that the unity of tempo remains intact.

The social, the political and the artistic implications of tactility could only have been lost to human awareness in a visual or civilized culture which is now dissolving under the impact of electric circuitry. The Japanese sense of the importance of touch as interval is sufficiently indicated in *The Book of Tea* by Okakura-Kakuzo (Kenkyusha):

> We must know the whole play in order to properly act our parts; the conception of totality must never be lost in that of the individual. This Laotse illustrates by his favourite metaphor of the Vacuum. He claimed that only in vacuum lay the truly essential. The reality of a room, for instance, was to be found in the vacant space enclosed by the roof and walls, not in the roof and walls themselves. The usefulness of a water pitcher dwelt in the emptiness where water might be put, not in the form of the pitcher or the material of which it was made. Vacuum is all potent because all containing. In vacuum alone motion becomes possible. One who could make of himself a vacuum into which others might freely enter would become master of all situations. The whole can always dominate the part.

The author continues his discussion in a passage that is perhaps better accommodated to Western perception:

> In art the importance of the same principle is illustrated by the value of suggestion. In leaving something unsaid the beholder is given a chance to complete the idea and thus a great masterpiece irresistibly rivets your attention until you seem to become actually a part of it. A vacuum is there for you to enter and fill up to the full measure of your aesthetic emotion.

II. A NOTE ON COLOR TV

Neither black and white nor color television is a picture. It is an X-ray. Light comes through the image at the viewer; the viewer is not a camera, but a screen. The TV camera has no shutter but works like a shifting mosaic. Totally different from photographs and movies, the TV image is discontinuous and flat. That is, it is a world of intervals. It is extremely tactile and participant.

The current phrase "living color" is self-contradictory since the TV industry takes "living" to mean realistic and representational. A similar mistake was made by the movie industry when it pushed toward photographic realism rather than realism of process, as in the Chaplin pictures. Chaplin was so conscious of this aspect of film that he frequently acted his scenes backward, although they were projected to run forward. Tony Schwartz has a tape on which Chaplin is speaking phrases backward, analytically and anatomically. Yet when played, they sound like normal speech. This is similar to the Stratton glasses which make us see the world upside down. The fact is that we do see the world upside down, righting it mentally for reasons that are totally unknown. The movie, because of its movement, strongly favors this awareness of processes in or out of nature. Photographic realism nullifies this aspect of the movie altogether. The talkie bypassed process in favor of simple narrative. TV is at present trying to recap all the worst mistakes in the history of movies. Since there are very few color experts in the world, there is very little hope of their having any influence on the television industry. In any event, color TV demands a considerable educational effort in perception.

We could say that the TV generation of teen-agers is, of course, getting the two-dimensional message of TV with all the force that the Negro community got the message of radio in the twenties. Naturally, the industry will continue to ignore the two-dimensional aspect of TV, and anyone over thirty is certainly not going to feel the new habit of perception latent in the TV image. It is precisely here that the artist has a crucial role to play in alerting human awareness to the meaning of technology.

The center or macula lutea of the eye is responsive to hue and variations in hue and texture. The periphery, on the other hand, is concerned with darkness and lightness and also with movement. Although the eye can eliminate any consideration of hue (for example, in scotoptic or twilight vision), when the macula is involved, it is inevitable that the periphery be involved as well. In other words, the macula and periphery work in tandem. However, peripheral vision can exist by itself. While color vision is inclusive, black and white is partial. Black and white TV is automatically inclined toward movement just as surely as color TV is inclined toward stasis and iconic values.

The potential of any technology is always dissipated by its users' involvement in its predecessor. The iconic thrust of color TV will be buried under mountains of old pictorial space.